# *The Age of Grandeur*
## *and a Woman Who Lived It*

### *Artist Evelyn Metzger*

# The Age of Grandeur and a Woman Who Lived It

# Artist
# Evelyn Metzger

BRETT TOPPING
NANCY G. HELLER

The National Museum of Women in the Arts
Washington D.C.
1995

## *Acknowledgments*

We would like to thank the young women who have worked as interns and volunteers in the museum's Publications Department over the past several years: Cristin Curry, Kate Kim, Holly Crider, Cynthia Graber, Kate Gibney, Anne McCaw and Jennifer Baggs. The hours they spent transcribing interviews, conducting research, locating photographs, checking facts and copyediting were vital to the development of this publication.

Designed by: Susan Rabin Design,
Baltimore, Maryland
Typeset by: BG Composition,
Baltimore, Maryland
Printed by: Balding + Mansell,
Peterborough, United Kingdom

LIBRARY OF CONGRESS CATALOGING-IN-PUBLICATION DATA

Topping, Brett, 1947-
    The age of grandeur and a woman who lived it: artist Evelyn Metzger / Brett Topping, Nancy G. Heller.
    p. cm.
    Includes bibliographical references.
    ISBN 0-940979-29-2 (hardcover). -
ISBN 0-940979-29-4 (softcover)
    1. Metzger, Evelyn Borchard, 1911- .
2. Women artists --United States
--Biography. I. Metzger, Evelyn Borchard, 1911- . II Heller, Nancy G. III. Title.
N6537.M485T66 1995
709'.2--DC20
[B]
                94-37120
                CIP

ISBN 0-940979-28-4 (softcover)
ISBN 0-940979-29-2 (hardcover)

© 1995 The National Museum of Women in the Arts, Washington, D.C. - compilation
© 1995 Evelyn Metzger - artwork and family photographs

Cover: Evelyn Borchard, a hand-colored photograph from 1934.

Page 6: The Borchards—Sam, Evelyn, Eva and Stuart—at La Solana, Palm Beach, ca. 1928.

Page 118: *Self-Portrait with Chandra*, 1991. Oil on panel, 35¼ × 27⅛ in.

## Contents

**6** *Evelyn Metzger: Life Impressions*
BRETT TOPPING
Notes
Bibliography

**118** *The Rich Artistry of Evelyn Metzger*
NANCY G. HELLER
Notes

**258** *Biographical Summary*

**260** *List of Art Reproductions*

# Evelyn Metzger: Life Impressions

BRETT TOPPING

**F**ORTUNATE to have been born into a family that always could afford the best of everything, Evelyn Borchard Metzger has traveled through life first class. As a child she rode grandly through the streets of New York in a chauffeur-driven Rolls-Royce, rolled her hoop on the decks of the *Île de France* and lunched with directors of the world's leading art museums. After graduating from Vassar she lived in a sumptuous apartment on the 33rd floor of the Waldorf Astoria Towers and vacationed at her family's homes in Palm Beach and Maine. As Wallace Brockway has written about Berthe Morisot, another woman artist born into exceedingly comfortable circumstances, "she might, but for her good sense, have easily been spoiled."

Not only is Evelyn level headed, she is wonderfully self-possessed. Her equanimity stems largely from her abiding devotion to art. The study and creation of art have been the determining factors of her life, providing its ballast, goal and direction. While the pursuit of artistic excellence has been a continuing source of stimulation, pleasure and self-esteem, it has increased the complexity of her life. As also was the case with Morisot, Evelyn's life-long commitment to art has required a refined ability to balance opposing interests and concerns—to find time to follow her own creative vision while fulfilling her extensive familial and social obligations.

Evelyn inherited her common sense and appreciation for art in equal measure from both sides of her family. Her father, Samuel Borchard, heralded the flow of Westerners that was to fill the ranks of New York's business and arts communities as the 19th century drew to a close. He was born in San Francisco in 1866 to an American mother (Elizabeth) and German father (Herman), but soon moved with them to New York, where his father sought greater opportunities.

1
Sam Borchard at two, 1868

2
Opposite: Eva Borchard in turn-of-the-century finery.

Growing up with a high regard for education (Herman Borchard was a professor), Samuel (known to family and friends as Sam) attended City College in New York from the age of fourteen. At this period he earned extra money by teaching math at night to college students who were considerably older. This age difference reportedly led to some unruliness in class, including at least one incident of erasers flying between teacher and students. The rigors of academic life, at both ends of the pedagogic exchange, seem to have amply prepared Sam Borchard for success in New York's business world. In the waning years of the century, he became an increasingly successful entrepreneur, enabling him to take on more responsibility for the support of his family and act as its de facto head. As a young man, his exercise regimen included runs around the Central Park Reservoir, during which he often sang "Ye crags and peaks I'm with thee once again!"

By the first decade of the 20th century, Sam owned S. Borchardt & Co. (the *t* was dropped from the name during World War I). The business occupied two sites—315–325 East 103d Street and 324–336 East 104th. Merchandise receipts provided to clients described the company's somewhat eclectic group of products: "Manufacturers of Overgaiters, Leggins, Lambs Wool Soles & Novelties." Eclectic or not, Sam's products sold.

One summer while vacationing at a resort near New York, Sam met Eva Rose. The youngest of thirteen children, Eva had grown up on a farm in Detroit. Her parents were close friends of James Garfield (dating back to the years before his election as president of the United States). In fact, Garfield attended their wedding and presented a handsome gift. Although Eva preferred to say that her father was in the candy trade, he was actually a tobacco importer. Given her beauty and effervescent self-assurance, it is not surprising that Sam was

3
Borchardt & Co. receipt, ca. 1905

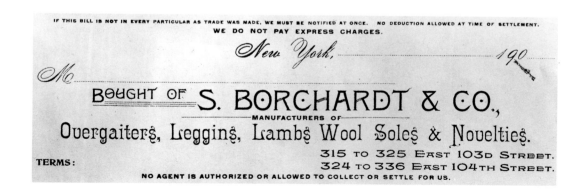

*4*
349 West 86th Street
(townhouse on right).

"*More gracious in style than most found on the East Side, [the townhouse was] built on the new American basement plan, so called because a wide, street-level reception hall replaced the traditional high stoop.*"

determined to wed Eva almost from the moment he met her. They married in 1902. Sam was thirty-six and Eva was little more than half his age.

Sam moved his young bride to his home at 349 West 86th, just off Riverside Drive, which he had purchased three years earlier. Built in 1890, the townhouse was "more gracious in style than most found on the East Side," as James Trager notes in *West of Fifth,* a social and architectural history of the upper West Side. "[It was] built on the new American basement plan, so called because a wide, street-level reception hall replaced the traditional high stoop." Where brownstones predominated on the East Side, the West offered a variety of façades, ranging from brick, like the Borchards' home, to marble, sandstone and limestone. Unusual rooflines and imaginative window treatments gave the neighborhood an "air of frivolity that made it seem almost a different city from the rest of town." An increasingly popular address, the West Side became a booming real estate market for wealthy New Yorkers. The average new row house cost $15,000 in 1890; by 1902 the cost had risen to $64,000.

## *The Opulent Promenade*

Turn-of-the-century photographs of Riverside Drive reveal a world of luxurious ease. A wide thoroughfare where mansions and exclusive, low-rise apartment houses confront a spectacular river view, the boulevard set a standard of European

5
Soldiers' and Sailors' Monument - Riverside Drive, by Irving Underhill.
The Underhill Collection, Museum of the City of New York.

6
Overleaf: Riverside Drive, ca. 1905. Collection Prints and Photographs Division, Library of Congress.

7
Eva Borchard

elegance and gentility for the entire Upper West Side. "I must bow before the majesty of such an avenue," enthused French novelist Pierre Loti with typical Gallic expansiveness.

The Borchards filled the early years of their marriage with travel, art collecting and the acquisition of real estate, setting the pattern for the decades to follow. Frequenting Europe's capitals, they trained their eye in the great museums and purchased their own art collection at renowned galleries in the United States and abroad. Sam's tastes were conservative, comfortably rooted in Old World tradition; Eva had more appreciation for the novel and daring. Together they assembled a collection which focused originally on 17th-century Dutch artists—Frans Hals, Meindert Hobbema, Jacob van Ruisdael, Pieter de Hooch, Jan Davidsz de Heem. In later years they expanded the collection to include Italian works by masters such as Giovanni Bellini and Bernardino Luini.

Sam also had an appreciation for fine jewels and enjoyed seeing Eva wear them. Pearls held a particular fascination for him. (These were the days when banker Morton F. Plant exchanged his Fifth Avenue mansion, a "splendid six-storey Renaissance palace" at 52nd street, for Pierre Cartier's most valuable two-strand pearl necklace.) When cultured pearls first appeared on the scene, Sam invited Mikimoto, who pioneered their development, to call at the Borchard's 86th Street home, where he explained the techniques used to achieve these luminous marvels.

Sam and Eva were familiar with the continental spas. A favorite was the delightful Bohemian resort of Marienbad, which Goethe found so congenial for its "magnificent quarters, civil landlords and good society." During one Marienbad interlude the Borchards were promenading through one of the spa's private parks, when gentlemen began handing Sam their entry tickets. Later in the day they learned that the men were voting for the loveliest lady there, and that Eva had won the beauty contest—the first foreigner to do so. Her prize was a magnificent ostrich-feather and tortoiseshell fan.

*8*
Evelyn Borchard, ca. 1915

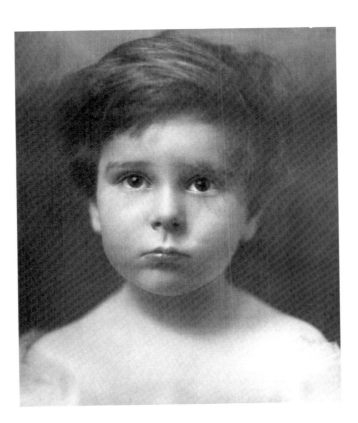

9
Riverside Park, the Hudson River and the Palisades, by A. L. Witteman. The Leonard Hassam Bogart Collection, Museum of the City of New York.

## A Cosmopolitan World View

When the children were born, Stuart in 1909 and Evelyn in 1911, their lives soon mirrored their parents'. From her first years Evelyn exhibited traits that have defined her throughout her life—passionate enthusiasm for art, adventurousness allied with a vigorous independent streak, keen intelligence and a strong sense of propriety.

Both children traveled to Europe from an early age, often staying with nursemaids at a hotel or resort while Sam and Eva went on museum and gallery excursions. Later, they began to accompany their parents on museum tours, forging a bond around art that was to be a mainstay of the family's relationship.

Evelyn's self-confident independence shines through her vivid memories of traveling by ocean liner as a child. "It was marvelous—Stuart and I would promptly explore a ship from top to bottom; we went right through it. Loved it. Everything was fun on steamers. I don't think anything gave us a bigger thrill." One of the activities that was particularly entertaining for the youthful Evelyn on board the liners was hoop rolling. She became so adept at it that she could not only pass through her hoop on the fly but also make it climb stairs. Being eager to make new friends and having good sea legs, she also remembers getting into

*The Royal Poinciana, "the grandest lady of them all," was one of the stars in the galaxy of hotels created by Henry Flagler.*

---

10
A Fair and Good Breeze -
The Vanderbilts and their
Friends at the Royal Poinciana,
1896. Collection Henry
Morrison Flagler Museum,
Palm Beach, Florida.

11
The palm-lined walkway connecting the Breakers Hotel and the Royal Poinciana, ca. 1900, where Evelyn learned to ride a bicycle. Collection Prints and Photographs Division, Library of Congress.

trouble for bothering older passengers seated on deck who were vainly trying to overcome the effects of seasickness.

Some of Evelyn's other recollections of ships during these years are less felicitous. As New York began to anticipate war in 1916, warships moored on the Hudson River off Riverside Drive became a more frequent sight. She recalls troops on their way to the front parading around the Soldiers' and Sailors' Monument on Riverside Drive while singing "Over There." She also remembers her father's strict admonition that she and Stuart should no longer speak German—the native language of their first governesses—because of mounting animosity toward the Germans. On a lighter note, when Edward, Prince of Wales (later the Duke of Windsor) visited New York at the end of 1919 to celebrate the war's conclusion on the first anniversary of the armistice, Evelyn recalls that her parents had the opportunity to meet him at a reception held at the Columbia Yacht Club at the foot of West 86th Street.

# *Luck Be a Lady*

When Evelyn speaks about her life, the word *luck* is the leitmotif. She mentions how very lucky she was that her brother, Stuart, was so fond of her throughout his life, despite the fact that her father often made remarks to him that easily could have led to resentment. "Why don't you have common sense, like your little sister?" was one of Sam's queries. She refers to her luck in having such a beautiful, warm, loving mother who cared so much about her and a father who, while less approachable and expressive, took a keen interest in his children. The word also characterizes her recollection of one of the most memorable trips of her youth—her first visit to Palm Beach when she was six.

The Borchards left New York in the throes of a freezing winter. Evelyn and Stuart, having been ill for weeks with whooping cough, had little energy for the trip. The journey took two nights and a day. When the train pulled up to the railroad arcade at the entrance to the Royal Poinciana in Palm Beach she was transported by the imposing yellow facade of the hotel, the radiant sunshine and a warmth she hadn't known for months. "It was like paradise."

*12*
*Coconut Grove Tea Dance*
ca. 1921
Ink on paper
5⅛ × 6½ in.

The Royal Poinciana, "the grandest lady of them all," opened on February 11, 1894. She was one of the stars in the galaxy of hotels created by Henry Flagler, the Standard Oil tycoon and railroad magnate whose vision transformed the Palm Beach coastline from mosquito-infested swamplands into the resort of choice for America's Gilded Age millionaires. The season opened casually in mid-December and ended abruptly on February 22 following the Washington's Birthday Ball, an event that Evelyn felt sure really honored her mother's birthday on the same day. During these two winter months each year the lavish private railroad cars of America's wealthiest men—Harry Payne Whitney's "Adios," Harry F. Sinclair's "Sinco," Joshua Cosden's "Roamer"—flanked the Royal Poinciana's entrances. It was the largest hotel in the world at the time, and the largest wooden structure ever constructed, graciously offering the height of luxury to 1,750 guests.

With the distractions of World War I over, Palm Beach enjoyed its "second great flourishing" in the 1920s. *Vanity Fair* editor Frank Crowninshield found the resort "not exclusive, but merry, sumptuous and expensive." From Evelyn's youthful perspective Palm Beach was memorable for the heavenly oranges that were peeled and stuck on forks for her by the white-clad waiters in the Poinciana's dining room. She recalls the tree-lined walkway between the Poinciana and the Breakers, where she learned to ride a bicycle. She was also impressed by the sophisticated tea dances that took place every afternoon at the Royal Poinciana's Coconut Grove, which inspired some of her earliest artwork. Evelyn's early sense of style is clear in her handling of this subject, which focuses on the dramatic picture hat worn by a swaying dancer in the foreground, balanced against the angular contours of the oriental gateway to the grove. "I was always drawing," says Evelyn of the winters she spent in Palm Beach. "Anyone who was more or less still for awhile, I would sketch."

---

*From her first years Evelyn exhibited traits that have defined her throughout her life— passionate enthusiasm for art, adventurousness allied with a vigorous independent streak, keen intelligence and a strong sense of propriety.*

She was not moved to draw a much more dramatic event, which she later experienced in Palm Beach after her parents had begun renting an apartment for the season—the fire of 1925, which demolished portions of both the Breakers and the Royal Poinciana. Evelyn still clearly remembers watching the terrified people who streamed toward the beach that night, loudly praying for their lives. Fortunately, the Borchards survived the ordeal unscathed.

## The Family Tree

Evelyn describes her parents' marriage as one in which two opposite personalities forged a strong union through accommodation, particularly on Eva's part. Sam was extremely proud of his wife's beauty, charm and ability to put anyone at ease. Throughout their married life he spent a great deal of time at her side. The Borchards were a good team. Sam drew constant pleasure from his wife's companionship; Eva felt secure being introduced to the wide worlds of travel, society and culture by a handsome, older man whom she respected and loved.

The conflicts which arose resulted from their natural human instincts. While Sam greatly admired Eva and was pleased that others found her so attractive, he was also jealous of their attentions. On her part, Eva, being naturally more vivacious and adventurous than her husband, sometimes felt restricted by his conservatism, his Old World sensibilities and his tendency toward prudishness. Being a woman of this era and sincerely devoted to her husband, Eva made most of the adjustments and concessions necessary to maintain harmony in their marriage. Sam also knew how to compromise, however.

One example of the dynamics of the Borchards' marriage relates to the fate of Borchardt & Co. As the century progressed, this successful enterprise required more and more of Sam's time. The hours he devoted to the company's management reached a point at which Eva began to feel that her husband was jeopardizing his health by working so hard. She would speak to him about cutting back, to no avail. Then one afternoon in 1917, when Sam was at home resting, the Borchards received a telephone call.

"I have some news for you, dear," said Eva.

"Yes?" responded her husband.

"Your factory is burning," came the staggering report.

It was completely destroyed. With the country entering World War I, Sam stood to make a fortune by rebuilding and supplying "leggins" and other items of clothing to the army, but Eva did not want him to. She was convinced that continuing to operate Borchardt & Co. would be detrimental to him. Bowing to his wife's wishes, Sam accepted the loss and shifted his capital and entrepreneurial skills to real estate.

## Ten at Last!

Turning ten is a watershed for many children. Fulcrum-like, this year separates the relatively unconscious pursuits of childhood from the first real awareness of self, the first strivings toward maturity. Evelyn's tenth birthday in June 1921 found her in Europe, where she had spent previous summers. What made this birthday different was the fact that she was in Dresden, where she saw Raphael's *Sistine Madonna* for the first time. What made Evelyn different and more culturally precocious than most other ten-year-olds was that the most important feature of this long-awaited event related to art.

Her parents marked her special day by giving her a delicious cake, covered with tiny macaroons, and the book *Stories for a Ten-Year-Old*. Evelyn also associates her tenth-birthday summer with Wilhelm von Bode, the internationally known director of the Kaiser Friedrich Museum, whom she was taken to meet in Berlin. An imposing figure, talented administrator and eminent scholar, Bode published extensively throughout his career on such wide-ranging subjects as Rembrandt's paintings, Italian Renaissance sculpture and oriental rugs. These publications were later to become Evelyn's textbooks at Vassar. Studying Bode's books in college thus became something of an ongoing dialogue about art, a dialogue which began the summer she was ten. "Everything was in context afterwards. When I read any of his books at college later on, I remembered having met him, which made it that much more interesting."

Some of the sights and experiences of the summer of 1921 were far less pleasant, making a lasting impression on the children. Evelyn clearly remembers the shock and pity she felt at seeing German war veterans, who took off bandages to expose the wounds they had suffered for their country, lining the streets of Berlin. The images lingered for days, stimulating a number of worried questions from both children about what had happened to these men. German marks were worth very little, a fact that even ten-year-old Evelyn noticed. As Senator Robert La Follette

13
Wilhelm Von Bode,
Director of the Kaiser
Friedrich Museum, Berlin.
Collection North Carolina
Department of Cultural
Resources.

of Wisconsin reported just two years later, as the situation worsened: "The Germans are suffering for want of food, fuel and clothing. Young children and old people are dying daily from hunger and disease. Emaciated, despairing, they are waiting for the end. Food riots are common. The crisis which is at hand involves possibilities too awful to contemplate. It menaces more than Germany."

## And the Band Played On

But Americans were in no mood to contemplate the gathering despair in Germany as the 1920s progressed. On this side of the Atlantic "something had to be done with all the nervous energy stored up and unexpended in the War," in the words of F. Scott Fitzgerald, the embodiment of that exuberant decade.

This was the decade of the newly independent woman, her hair cut in a radical "boyish bob." Its insignia—fast cars, bootleg gin, raccoon coats, the Charleston and jazz—honored the era's twin gods of pleasure and speed. The decade "raced along under its own power, served by great filling stations full of money," reminisced Fitzgerald.

14
*Deco Style*
ca. 1921
Ink on paper
5½ × 3½ in.

15
*Triangle Lady*
ca. 1921
Ink on paper
5 × 3 in.

*16*
*Young Sophisticate*
ca. 1921
Ink on paper
5⅛ × 3⅝ in.

*17*
William R. Valentiner in Berlin, 1919. Collection North Carolina Department of Cultural Resources.

*As teenagers both Stuart and Evelyn regularly attended their parents' Sunday dinners. Conducted in an atmosphere of cultured ease, the dinners introduced Evelyn to a variety of interesting and highly prominent members of the art world.*

New York was the place to be in the '20s, and it seems that everyone had heeded its siren call: Damon Runyon, a reporter with an ear attuned to the colorful cadences of Broadway; the Algonquin Round Table—Dorothy Parker, Alexander Woollcott, Neysa McMein, George Kaufman and friends—who were to become famous for their conversation and sarcastic repartee; songwriters Irving Berlin and Cole Porter; stars of the nascent film industry—Lillian Gish, Tallulah Bankhead, Charlie Chaplin.

An episode which encapsulates the zaniness and freedom of the Jazz Age was recounted by Jack Baragwanath, husband of the famous *Saturday Evening Post* cover illustrator Neysa McMein. It took place at a dinner party in the studio of sculptor Sally Farnham, who later taught sculpture to Evelyn. Early in the evening, Sally's pet monkey was making a bit of commotion, so Farnham handed him a crêpe suzette. The monkey draped the dessert over his shoulders and soon quieted down, soothed by its warmth. "Evidently it was the usual remedy, for no one else seemed to take any notice," remarks Brian Gallagher in his lively portrait of the 1920s, *Anything Goes*.

Evelyn studied with Farnham in the early 1930s, learning modeling techniques that enabled her to create extremely realistic portrait busts of her family and friends. Although she clearly remembers meeting Neysa McMein, whom she found to be exceedingly attractive, at Farnham's studio, the pet monkey was no longer in evidence ten years later.

## The Art of Life

In the face of the frantic hedonism which characterized the '20s, the Borchards maintained a slower pace, pursuing their own quiet pleasures. One of these was art, both viewing and collecting it. Their regular trips abroad revolved around visits to museums and to art dealers. They eagerly shared this interest with their children. "I feel privileged that, starting at a very young age, I was exposed to the best in art," recalls Evelyn. "My father was the intellectual. He would go to the library and study in museums," adds Evelyn. "My mother felt things. She'd look at something and decide this was wonderful, and she had excellent taste. Between the two of them they did very well; they really didn't make too many mistakes." Evelyn inherited her broad appreciation for originality in art from her mother and her intellectual curiosity, which helped her to shine later in her art history courses, from her father. From an early age Evelyn shared Eva's admiration for certain works. "I remember our favorite area in the Louvre was alongside the large Rubens rooms—little galleries housing the van Eyck, the Pieter Brueghel, a Fouquet, the Avignon Pietà. Superb, early paintings—14th-, 15th-century. They mean much more to me than the huge Rubens."

When asked about her early art education, Evelyn explains that nearly all of it took place within the family circle, through visits to museums and conversations with family friends. As teenagers both Stuart and Evelyn regularly attended their parents' Sunday dinners. Conducted in an atmosphere of cultured ease, the dinners introduced Evelyn to a variety of interesting and highly prominent members of the art world, nurturing the profound love of art that her parents' own enthusiasm had already established. Sunday dinner guests included the well-known art dealer Joseph Duveen. "He received phenomenal prices from J. P. Morgan, Wiedener and others," notes Evelyn. "With the help of Bernard Berenson, Duveen put 15th- and 16th-century Italian painting on the artistic map."

Of all of her parents' luncheon guests, Evelyn was particularly taken with the cosmopolitan and inspiring art historian William Valentiner. One of Bode's protégés, Valentiner was appointed Curator of Decorative Arts at New York's Metropolitan Museum in 1908 and later became the "father" of the Detroit Institute of Arts. "Valentiner's great quality as an art historian and expert was his instantaneous and enthusiastic reaction to beauty," notes renowned art dealer Germain Seligman in assessing his remarkably successful career. Although Valentiner's principal contributions to scholarship were in the fields of Italian Renaissance sculpture, 17th-century Dutch painting and German Expressionist art, he was, in fact, passionately committed to every period of art, from its beginnings to its most contemporary manifestations.

Evelyn remembers that because her father followed the letter of the law, these wonderful and stimulating Sunday dinners never included alcoholic beverages during the Prohibition years. Her mother did her best to compensate for this shortcoming through attention to all aspects of the meal, from the exquisite appetizers preceding it to the small gifts of silver and enamel, purchased on the rue de Rivoli in Paris during summer trips abroad, which each guest took home. "Mother loved to entertain," remembers Evelyn. "Her dinners were always events, with superb food."

In addition to the art world luminaries whom they entertained in their home, many of the Borchards' friends, particularly Eva's good friend Jane Peterson, shared their commitment to the arts. Another transplanted mid-Westerner, Peterson arrived in New York from Elgin, Illinois, in 1895 to enroll in the Pratt

18
Jane Peterson on the
SS *Majestic,* returning from
a painting trip to Egypt
in February 1934.
UPI/Bettmann Newsphotos.

Institute. Her worldly possessions consisted of $300, her secondhand clothes and a strong determination to succeed. A vivid, courageous and beautiful woman, Peterson quickly made important social contacts that would last her a lifetime, among them the financier Alexander Hudnut and designer Louis Comfort Tiffany. She completed her studies at Pratt and went on to Paris and "La Vie Bohème" of 1908, highlighted by Gertrude Stein's Saturday soirées and the Fauvist revolution. An early world-class traveler and free spirit, Peterson was soon packing her paint boxes, easels, supplies and equipment on unaccompanied trips through Egypt and Algeria. The paintings that resulted are "sparkling, sundrenched *plein-aire* impressions by an artist who was in love with life and determined that life would love her," in the words of Jonathan Joseph.

The friendship between Eva Borchard and Jane Peterson began in Palm Beach, where Peterson helped establish the Palm Beach Art League in 1918. They saw each other frequently in New York, exchanging visits between the Borchard's home on West 86th Street and Peterson's on West 57th. Following her marriage to the wealthy corporate lawyer M. Bernard Philipp in 1925, Peterson moved to a five-story townhouse on Fifth Avenue, directly across from the Metropolitan Museum of Art. The couple soon added another floor to create a magnificent artist's studio. It was in this house that Evelyn best remembers Jane Peterson, frequently associating the artist with the museum that has played such a central role in her own life. She also remembers painting beachscapes with Peterson in Palm Beach.

Others of Eva's friends also were very much immersed in the art world. Her best friend was Hattie Jonas, who lived in an art-filled duplex at 998 Fifth Avenue. The bedroom floor was reserved for van Goghs and Impressionist paintings, while

*19*
A Rolls-Royce Riviera Town Car, similar in style and finish to the model owned by Sam Borchard. From *Rolls-Royce in America* by John Webb de Campi.

Old Masters hung on the floor below. Although the seed of the collection preceded Hattie's marriage to the Parisian art dealer and real estate magnate Edouard Jonas, together they added many superb works through the years. Outstanding in its breadth and quality, their collection was a mecca for Evelyn and Stuart.

## Horace Mann High School

"Horace Mann School has for its function the demonstration of all that is best in teaching and school management under conditions that are as nearly ideal as possible." Such was the lofty goal of the institution in 1902, even as it had been in 1887, when Nicholas Murray Butler founded it as an experimental arm of Teachers College, Columbia University. And such the goal remained in 1925 when Evelyn entered the prestigious high school as a freshman.

From its inception, Horace Mann provided an education that was rigorous, demanding and progressive. The girls school, remaining at 120th and Broadway after the boys school split off in 1914, continued to adhere to the founder's progressive philosophy of education. One instance of this was the "project method" instituted by history teacher Roy Hatch in the 1920s, whereby students chose a topic for the year which became the focus of their research and presentations. By the 1930s the school had a fully interdisciplinary curriculum which grouped subjects historically, beginning with pre-history in the seventh grade and ending with contemporary times in senior year.

"The school taught students how to think, write and organize. It provided a marvelous preparation for later endeavors," reminisces Evelyn of these years. She recalls that in her day students went from classroom to classroom, as in college, to attend sessions with teachers such as Helen Baker. A stimulating English teacher who loved her subject, Baker nonetheless counseled students to take advantage of their college years to explore subjects other than English—anthropology or archaeology perhaps.

Responding positively to her father's own love of learning, Evelyn became an *A* student—serious, hard working and popular with her teachers. French was one of her strong suits, due in part to the numerous summers she had spent on the continent. During her senior year Evelyn was expected to win the citywide French medal. That year, however, Horace Mann students became ineligible for the medal, because private institutions were no longer allowed to compete.

Demonstrating their regard for Evelyn's scholarship, the school's French teachers responded by collectively awarding her their own prize—an inscribed book of French poetry.

After-school activities are among Evelyn's most vivid memories of high school. Sam frequently picked her up from class in his chauffeured Rolls-Royce with the snappy "basketweave" finish, and they would go up to Van Cortland Park to play golf together in the Polo Grounds. He would first demonstrate the proper technique. After awhile he would drop a ball saying, "All right, young lady, let's see how many strokes it takes you to hit this." In the winter Sam followed the practice of Rudyard Kipling, credited as the inventor of "snow golf," and had the balls painted red. The result of these afternoons together was Evelyn's powerful, even swing. In later years her swing made her a strong competitor against men and women, earning her shelves of trophies.

During her formative years Evelyn shared many of her father's favorite pastimes, along with his interests and opinions. Although she politely refrains from making such a direct statement, she clearly fascinated her father and was a source of great pride for him. "He was very fond of me," she admits, "though he had wanted only sons."

## The Young Sophisticate

By her mid-teens Evelyn's figure was graceful and willowy; her face delicate, with chestnut hair framing hazel eyes and classical features. In 1926 Evelyn and her family attended the wedding of Georges Clemenceau, grandson and namesake of the famous French statesman. The event took place at the bride's home, the exquisite Château de Louveciennes—frequently painted by Camille Pissarro and other Impressionists—near Paris. Evelyn remembers that "Le Tigre," as the elder Clemenceau was known, spoke to her at some length about his warm feelings for the United States. She also has a vivid recollection of the bitterness expressed by the mother of the groom, who was divorced from Clemenceau's son, about the lack of status and respect accorded divorced women in France. Evelyn, who looked quite sophisticated that day in a light blue dress and velvet picture hat, later found herself the focus of attention for several matchmaking mothers at the reception. Fortunately, Eva soon intervened. "Please, she's a bit young for this discussion," she said, dashing their fantasies of securing a wealthy American bride for one of their sons with the news that her daughter was still only fifteen.

A few years later jeweler Harry Winston lunched with her family at the Ritz in Paris. He struck Evelyn as something of an eccentric and as a person who loved a good prank. This impression was due in part to the fact that, when Winston shook her hand, he deposited a huge and magnificent diamond in it, causing her heart to skip a beat. Unfortunately, he later retrieved it. Another fact that caused Evelyn to feel that this king of the diamond trade had a very original way of doing business was his remark that he employed the U.S. mail—unregistered and uninsured—to ship the fabulous diamonds and jewelry that comprised his multi-million dollar enterprise.

Also staying at the Ritz on this trip were Ernest and Adelaide Makower, heirs to Makower and Co., Silk Merchants of Old Change, and prominent English art patrons. Learning that they had many common interests, the Makowers promptly invited the Borchard family to be weekend guests at their country estate, Little Binfield House at Henley-on-Thames. Among the other guests that weekend were Lord Cecil William Norton Rathcreedon and Sir Robert Filmer, a "land poor" owner of thousands of acres whom Evelyn remembers particularly for his badly worn tuxedo. While giving a tour of their estate, the Makowers, who were renowned antiquarians, proudly showed their guests their fine collection of antique china. Later, when the conversation turned to Elizabeth I's ring which they had recently given to the reigning queen, Ernest and Adelaide offered to arrange a presentation at court for Evelyn. Having more serious interests than making a splash as a young debutante, she politely demurred. "My mother was very smart," says Evelyn concluding this story. "She never forced me to do anything social that I didn't want to."

## Vassar College

A graph of Evelyn's emotions during her college years would reveal more than the usual peaks and valleys of feeling to which teenagers on the cusp of adulthood are susceptible. The years that Evelyn spent at college were tumultuous on many levels, for the country and for the family.

Evelyn entered Vassar in September of 1928. Attending this school was something of a tradition for the women of her family; Evelyn was preceded there by at least two of her mother's cousins. There was a bit of a tug-of-war between her parents before the decision became final, however, since Sam wanted his daughter to continue her education closer to home. Eva, fearing that her husband's conservatism might stifle her daughter's development, felt Evelyn should experience life away from home for awhile. Describing herself at this juncture as a very cooperative young woman, Evelyn was prepared to follow whatever decision her parents made. "I wasn't a rebel. I went along with what my parents decided, pretty much." In this case Eva prevailed, so Vassar it was.

Life at college turned out to be a mixed blessing. On the one hand Evelyn met a group of women who became the basis of a network of friends with whom she has been in contact ever since. Her conversation is peppered with references to women she knew at Vassar, or whom she met through Vassar. The Vassar connection provided the spark for her friendship with Jane Carey (class of '20), political advisor and friend to Mohammed Reza Pahlevi, Shah of Iran. Carey donated to the Metropolitan a splendid set of innovative Mary Cassatt drypoint etchings inspired by Japanese woodcuts. Other friends with ties to the college included Futurist art collector Lydia Winston; novelist Suzette Telenga; Gladys Delmas, a benefactor of the New York Public Library; and Evelyn's sponsors at the Cosmopolitan Club, Kay Shepard and Peggy Hyde. In spite of new and prospective friends, however, Evelyn was away from home for the first time and felt lonely. During her first year at Vassar, she went home as often as she could and relied on frequent telephone calls and chocolate cakes from her parents to sustain her between long weekends and vacations.

As always, when she was home, art became a primary topic of discussion and a focus of family activities. The Borchards and others in New York's art community had a great deal to say at this time when Sam succeeded in acquiring a jewel-like village dance scene by Pieter Brueghel the Elder—one of the thirty or so extant works by this incredible artist. This acquisition was a recognized coup, which further enhanced the Borchard collection. Sam was convinced that another work in the collection, *Christ among the Doctors,* was by Leonardo da Vinci. The Borchards felt that this version was superior to a painting of the same subject by Bernardino Luini in the National Gallery in London. "Father was so excited about his discovery that he had an aunt of mine take the painting over to Berlin and show it to Bode," says Evelyn. Bode wrote back that he thought it was, in fact, a Leonardo which had been in the Pamphili Palace in Rome for two hundred years. Then, of course, there was Giovanni Bellini's spectacular *Presentation,* a Madonna and Child shown with saints and a donor. This work was so beautiful and moving that it inspired Evelyn to write an article about it for the *Vassar Journal of Undergraduate Studies.*

Another topic of conversation on visits home was the ongoing search for new residences. "We've just bought two houses!" was the exciting news coming over the

*20*
Lytton Lodge, Maine

*21*
Evelyn and Stuart vacationing at Lytton Lodge, Maine, ca. 1928.

22
La Solana, Palm Beach

23
La Solana, Palm Beach

*"It was the happiest house
I've ever known."*

24
La Solana, Palm Beach

*Plate 1*
*La Solana*
ca. 1928
Oil on tin
4¾ × 12 in.

wires one evening in the first months of her freshman year. After many summers spent seeking the ideal French château, a living situation that appealed more to Sam than to Eva, the Borchards had settled on La Solana, one of Addison Mizner's Moorish-Venetian confections in Palm Beach. The house, built for Harold Vanderbilt but purchased from the Litt family of Philadelphia, came with a hunting camp in Maine, Lytton Lodge, thrown in.

"I never design a house," Mizner is quoted as saying, "without imagining some sort of romance in connection with it." While it is not known what specific love story the architect had in mind with regard to La Solana, the house he designed certainly had romantic allure. Just big enough without being unwieldy, it was adorned with Spanish tile floors and Moorish arches throughout. Colored Venetian glass bejeweled the windows; the topaz shades reserved for the sun room readily seduced occupants into a languorous, saffron-toned reverie. A wonderful music room, lovely 18th-century Spanish paintings, shaded patios and lush gardens completed the luxurious surroundings. "It was the happiest house I've ever known," recalls Evelyn with a wistful sigh. Something of La Solana's happy aura is evident in a small painting of the house which she completed in college. Demonstrating the adventurous spirit and love of experimentation which has defined her art throughout her career, Evelyn painted five views of the house—on the sides and top of a tin box. Not only was La Solana romantic, but Evelyn's life was blossoming with romance during those years. She remembers entertaining her first boyfriends on bicycle rides along Lake Drive and the ocean front.

Evelyn's feet came back to earth with a thud during her second year of college. A month into her sophomore year at Vassar the stock market crashed, plunging the nation into the worst economic crisis in its history. The crash brought an abrupt and stunning end to the excesses of the Jazz Age, which "as if reluctant to die outmoded in its bed, leaped to a spectacular death in October, 1929," wrote Fitzgerald in his poignant eulogy for the decade that made him.

Since Sam's investments were mainly in New York real estate, his energies having been directed toward building and acquiring properties following his years as a manufacturer, the family was spared the worst of the stock market's crash. In 1929 he owned fifteen buildings, including an apartment house on Park Avenue which he built in 1916 and jokingly offered to Evelyn as a future wedding present. As the Depression worsened, the family was forced to drop several properties, including 32 Broadway, a prime office building in the financial district. Like other real estate owners, they had tenants who had lost their assets or their jobs, and could no longer pay their rents. This meant that rent revenues no longer

*Plate 2*
*Sam Borchard as a Young Man*
n.d.
Oil on canvas
36½ × 23½ in.

*Plate 3*
*Stuart Borchard*
ca. 1928
Graphite on paper
16 × 12 in.

supported the mortgages or taxes on a building. Despite the blow dealt them by the crash, the family continued to live very well.

A much more devastating blow to the family was caused by Sam Borchard's death in March of 1930 at La Solana in Palm Beach. Their shared loss forged an even closer union among Eva, Stuart and Evelyn. Twenty-year-old Stuart assumed major responsibility for managing the family properties. Due to their illiquidity for decades to follow and the advent of residential rent control, management of the properties developed into a burden he would resolutely bear all of his life. Evelyn became more actively involved with the family art collection, obtaining additional "expertizations" and documentation, as well as adding works. At this period she spent hours doing research at the Frick Museum library, which already constituted an impressive resource on the history of art. The Frick archives grew out of a project initiated by Helen Frick, daughter of steel tycoon Henry Clay Frick, to document art in private collections throughout the country. In the 1920s Helen had taken photographers to the Borchard's home to record their artwork, so Evelyn had known the library really from its infancy.

*25*
815 Park Avenue, built by Sam Borchard in 1916.

26

220 West 98th Street—
The Borchard—was built by
Sam Borchard in 1911.

27

The Borchard's lobby
was filled with Tiffany
glass panels.

A Vassar story that Evelyn tells on herself to illustrate a light-hearted forgetfulness actually reveals information about her increased role in managing the family collection. During her junior year the art faculty held a tea for a distinguished guest lecturer, William Valentiner. The student who was organizing the tea, Connie Haass, asked Evelyn to help her, knowing that Valentiner was a family friend. On the afternoon of the reception, Connie and Evelyn greeted Valentiner, welcomed him to the school and began chatting. The next thing they knew the hour for the tea had passed, and they had forgotten to take him there. Valentiner

*28*
The sleek twin towers of the Waldorf Astoria dominated the New York skyline in 1931. Collection Museum of the City of New York.

29
The elegant living room of the Borchards' apartment 33H in the Waldorf Towers.

was delighted at having spent such a gay afternoon in the company of two charming young ladies instead of having tea with the faculty, and all ended well. That weekend, she adds almost as an aside, Valentiner returned to New York with her to spend several days studying the family's 17th-century Dutch paintings.

Two years later, in June 1932, Evelyn was Connie's maid of honor at her Grosse Pointe home for her wedding to Trent McMath. Among the guests was Henry Ford, Sr., who spent a good part of the festive occasion discussing Christian Science with Eva.

Another event that marked Evelyn's junior year was moving with her mother and Stuart into apartment 33H of the Waldorf Astoria Towers. This change of residence was prompted by Eva's reluctance to return to West 86th Street after Sam's death. The move also reflects what Lloyd Morris terms the "predilection for diminished responsibility and greater convenience" which accounted for the increasing popularity of luxury apartments over private residences among New York's upper classes in the first decades of this century.

And a luxurious existence it was. Completed in 1931, the Waldorf Towers offered magnificent views of New York from the Borchard's apartment. Elegantly proportioned, the floorplan encompassed a living room, dining room, library and three bedrooms—each with its own exquisite marbled-and-mirrored dressing room and bath—in addition to the kitchen and maid's room. The family had their own maid to attend to their needs, and enjoyed the full services of the Waldorf Astoria Hotel. "If anyone took a nap in the afternoon, the sheets would be changed," recalls Evelyn, "and the bathrooms had bars of the finest imported soap that would be replaced once they were used a few times."

Among the first tenants to move into the Waldorf Towers, the Borchards had the opportunity to look at every apartment and select the one they liked best. In short order the Waldorf Towers became one of the most glamorous addresses in New York, patronized by presidents, emperors and film stars. "We'd run into the Duke and Duchess of Windsor in the elevator," remarks Evelyn. "Herbert Hoover and his family lived there, as did Cole Porter. Emperor Haile Selassie of Ethiopia visited frequently. The Waldorf was known as the palace of New York."

## The Rain in Spain Stays Mainly in . . . Santiago de Compostela

One of her major influences at Vassar was art historian Agnes Rindge. Having been exposed to art continuously, Evelyn felt that she could coast through the introductory art history course. Prof. Rindge then took her down a peg by giving her a *C* in this "baby art" class, as Evelyn calls it. After that, Evelyn settled down to learn as much as she could from her stimulating teacher, characterized recently by feminist art historian Linda Nochlin as "elegant, worldly and sophisticated, but also a bit wicked." After four years of college, Evelyn graduated from Vassar Phi Beta Kappa in 1932. Having completed a course on Spanish art with Rindge in her last semester, which focused mainly on the country's architecture and devotional objects, rather than painting and sculpture, Evelyn was eager to tour

# "We'd run into the Duke and Duchess of Windsor in the elevator."

the Iberian peninsula with her mother the summer after she graduated. In the early 1930s Spain was still virtually unexplored territory for Americans and Europeans alike. It is a tribute to Evelyn's natural adventurousness and Eva's love of the unusual that they would undertake this excursion so far off the beaten path.

Evelyn's eager anticipation of the trip increased when she visited theaters and cultural centers in Spanish Harlem with artist Malvina Hoffman, who had recently returned from South America, where she had been researching her famed series of ethnographic sculptures produced for Chicago's Field Museum. The two women greatly enjoyed spending time together in the weeks before Evelyn's departure—discussing sculpture, visiting museums and immersing themselves in the exoticism of Harlem's Latin communities.

Arriving on the continent, the Borchard ladies traveled to Spain in style—driving down from Paris in a handsome, extremely commodious Minerva touring car, accompanied by a chauffeur and maid. Evelyn reports that their diet consisted largely of *churros* (sugared crullers) and hot chocolate, since dinner was served after 11:00 P.M., too late for early rising sightseers. On one occasion, however, in San Sebastián, their maid, who had lived in Spain, suggested they have tea in what appeared to be a particularly elegant restaurant. Feeling ravenous, the three women ordered lavishly. When they asked for their bill, they were embarrassed to learn that they were actually having tea in an exclusive private club. Their unease was alleviated shortly, however, when the Conde de Güell, a valued club member, came to their table and assured them he was "honored to have had the opportunity to extend his club's hospitality to such distinguished guests." The friendly politeness of the Spaniards, displayed on this and numerous other occasions, was a constant source of pleasure for them.

The incredible art in Spain, previously experienced only through slides and reproductions, provided spiritual sustenance on the trip. Evelyn vividly remembers visiting the sights she had recently learned about in class, "Visigothic churches, little churches, around the west coast and in the north. You would have to go down in the village and get someone to give you a key to open some of the sanctuaries. We also visited the south, which is filled with beautiful Moorish art." Santiago de Compostela was one of her favorite sites. "It has later buildings—Gothic and early Renaissance—but the Romanesque churches and sculpture are the most magnificent. The whole town has colonnade-covered streets protected from the rain. It's really a Romanesque town, very beautiful, very rainy, very near Portugal. The language they spoke was nearer Portuguese than Spanish."

The art treasures of Spain were so far removed from the tourist circuit in those days that, in most cases, Evelyn, Eva and their maid were the only visitors at each destination. At the Colegio de Santillana del Mar near Oviedo, Evelyn held the 11th-century *Prayer Book of Fernando and Sancha* (1055), created for Ferdinand I, king of Castile, and tremulously turned its vellum pages. When they visited Altamira the three women followed a guide into the caves and were surrounded by the transcendental presence of the Paleolithic. They stood marveling at the cave paintings with their ravishing colors—brilliant reds, yellows and blacks—and wonderfully expressive imagery. "Part of the magic was that we were alone, this was no organized tour." Because of the damage wrought on the fragile paintings by the breath of visitors and soot of smoking torches, the murals have darkened and the caves are no longer accessible to the public. Now tourists must content themselves with re-creations of the caves at the archaeological museum in Madrid. Evelyn says that the re-created caves represent murals that had already faded a great deal. "I don't suppose very many people alive today know how brilliant the colors were in those extraordinary 20,000-year-old paintings." This, like so many of Evelyn's experiences of art, can never again be duplicated.

The travelers arrived in Seville anticipating flamenco bravado, their ears attuned to the staccato of castanets. Eva and Evelyn asked their maid to consult the hotel concierge regarding where they might find the best dancers. After dinner that evening, the three women set forth and were soon being ushered to the back of a nightclub filled with men at small tables facing a stage. Their entry provoked quizzical stares all around.

Shortly after the women were settled at their table, the curtains parted to reveal the first performer—a heavy, naked woman who sang a number of bawdy ballads, judging from the audience's reaction. Not exactly what they had envisioned for the evening's entertainment! Several performers of equivalent dress (or undress) and repertoire followed. As it became evident that flamenco was not on the night's program, Eva, Evelyn and their maid rose to leave. All of the men in the room were perfectly polite. They moved their chairs to let the women pass, bowing them out without even the hint of a snicker or sneer. Although the travelers' pleasurable anticipation of witnessing a perfect Sevillian flamenco was dimmed by this episode, their appreciation of Spanish courtesy and gallantry was further enhanced.

When they had finished their tour of Spain, Evelyn and her mother returned to Paris for an extended stay. During this Paris sojourn they visited the Russian dealer Konstantinoff on the West Bank, where they purchased a group of Old Master drawings. These included a Rembrandt, a school of Michelangelo (known as Michelangelo in the Anatole France collection), a Lancret, a pair of Gillots, a Boucher and an Odilon Redon, among others.

Once back in the United States, they found that the economic impact of the Depression provided further buying opportunities. Evelyn and Stuart consulted with gallery owners such as Jackson Higgs, receiving a free education about the antiquities market. Evelyn is still amazed by how willing Higgs and others were to spend hours talking with two youngsters in their early twenties. They found the auctions at the Anderson Galleries—the predecessor of Parke-Bernet and Sotheby's—to be a particularly good source for many of the early Egyptian figurines, Roman-Syrian glass objects, medieval ivories and Persian ceramics which the family added to the collection at this time.

One memorable purchase was a Chinese sword from the Han period (206 B.C.–A.D. 221) with an inscription which showed that it belonged to the leading general of the day. Higgs spoke to Evelyn and Stuart at some length about the rarity and historic significance of the work, which made its purchase at auction even more exciting. It is now in the collection of the Kansas City Museum.

Among other artworks added to the collection in these years were a graceful Madonna and Child by Filippino Lippi and another by Andrea Vanni, both from Alfred Sambon of Paris. The Borchards also purchased a 16th-century Jean Pénicaud enamel of Jesus' descent from the cross from the sale of the Thomas Fortune Ryan collection in 1933, and bought a Pietà by Francia and a Madonna by Piero di Cosimo from Jackson Higgs.

## "I See a Tall, Handsome Stranger..."

Evelyn's life changed forever on a night in 1933 when she met Herman ("Ham") Metzger at one of her mother's parties at the Waldorf. As the festivities got under way, Evelyn's date for the evening approached her saying, "You must talk to that young Latin American, he's fascinating!" Intrigued, Evelyn struck up a conversation. She soon learned that this tall, energetic, captivating young man was not Latin, although his moustache suggested it. He was a Cornell-trained engineer working in the oil fields of Colombia. The couple had an instant rapport, and Ham soon asked Evelyn to dine with him later that evening. She had a previous engagement, but they went out the next three nights in a row.

Two people close to Evelyn had premonitions about her and Ham. Jane Peterson, who had known Ham for years, had been talking about getting the two young people together for some time. Her mother also foresaw a positive outcome for their friendship. It was their custom that whenever Evelyn returned from a date she would go to her mother's room and talk about it. After her first evening out with Ham, Eva said, "You're going to marry that Mr. Metzger and you're going off to South America. We can talk about it in the morning."

"But that's ridiculous," insisted Evelyn, "I've just met him!" With a mother's instinct, Eva sensed before Evelyn did the breadth of Ham's appeal and the seriousness of his intentions.

Born on July 1, 1900, Ham Metzger liked to say he was born on the fiscal New Year of the century. His father, Sylvain Metzger, had arrived in New York from Strasbourg, France, in 1887 at the age of thirteen. His mother, Tillie Block, was born in New York in 1876.

30
Ham Metzger (left) and James Flanagan holding a strategy session in the mid-1920s. A demanding friend, Captain Flanagan expected complete loyalty and devotion from younger associates.

From his early years Ham had impressed others with his charm, knowledge and enthusiasm. At the age of thirteen, he built his first radio and presented it to James Wainwright Flanagan, a family friend who would soon become a special investigator and agent for one of the most talented international businessmen of the 20th century—Walter C. Teagle, president of the Standard Oil Company (New Jersey). Ham entered Cornell's Sibley College of Engineering in 1917 on two scholarships. Capable of existing on only four or five hours of sleep, he was extremely successful at Cornell in both the academic and social spheres. He taught freshman physics as a sophomore, became a member of Phi Kappa Phi honor society and won the Sibley Prize both his junior and senior years. He enjoyed talking with friends on a wide variety of topics until the wee hours, after which he would stay up studying. When he graduated from Cornell in 1921 he had earned degrees in both mechanical and electrical engineering.

Prior to graduation Ham enrolled in the Officers Training Corps intending to go abroad and fight for his country in World War I. He was greatly disappointed when the war ended before he could join the soldiers at the front. Following graduation the wide world continued to beckon, and Ham's friendship with Flanagan soon resulted in a job that offered both experience and adventure—working in Colombia for the Tropical Oil Company, a newly formed subsidiary of Jersey Standard.

When Ham Metzger joined Tropical Oil in 1921, he began to learn the oil business, literally, from the ground up. Journeying on a United Fruit Line freighter from New York to Cartagena and then by steamboat 350 miles up the Magdalena River to Barranca-Bermeja, the nearest landing point to Tropical's drilling sites, he taught himself civil engineering and Spanish. He and his fellow workers spent the next three years laying the groundwork for a successful oil drilling and refining concession, working in singularly hostile and inaccessible terrain. "The De Mares concession itself was a wilderness—a land of steaming temperatures, unbelievable rainfalls, and none-too-friendly native tribes," notes George Gibb in his history of Standard Oil between 1911 and 1927. "Transportation facilities consisted of river boats and canoes, and the caprices of the Magdalena made navigation difficult." In these years Ham also participated in a six-months' jungle survey for the Andian National Corporation, a Canadian-based subsidiary of Standard Oil organized and operated by James Flanagan. The survey preceded the construction of the 1,000-mile Andian Pipeline connecting the oil fields to the Colombian coast.

During his first years with Tropical Oil, Ham was to display a wide range of skills and qualities that made him an invaluable addition to the Standard Oil family throughout his career. He proved to be a capable administrator, and was

*31*
Near left: Most travelers took the slow route up the Magdalena River to Barranca-Bermeja by riverboat.

*32*
Far left: Barranca-Bermeja drilling site in Colombia following years of hard work, ca. 1935.

promoted to construction manager and then to assistant general manager. By 1925 he was in charge of 5,000 men. He was also a popular companion, frequently called upon to play the piano for the workers' evening gatherings. During the survey for the Andian Pipeline, Ham became the group's pastry cook. Armed with the *Monmouth Baptist Mothers' Cookbook* and an oil-drum stove, he turned out many welcomed treats. His specialty was coconut cream pie. Since the survey team had decided to take along no alcoholic drinks, sweets were in great demand.

In 1928 Ham transferred to Bogotá as an executive, representing Tropical Oil and Andian National. Prefiguring later years with Jersey, his job was largely legal and diplomatic. He soon demonstrated his acumen as a negotiator by guiding Jersey lawyers to a successful conclusion of the companies' lawsuit against the Colombian government.

In Bogotá he developed close ties with other influential Americans residing in the capital. Among his best friends was his neighbor, Jefferson Caffery, minister of the United States legation in Colombia. One of the most successful foreign service officers of his day, Caffery was to receive increasingly important ambassadorial assignments—Cuba in 1934, Brazil in 1937, first career ambassador to France in 1944 and Egypt in 1949. The two men shared an enthusiasm for mountain climbing and spent many a Saturday climbing nearby Monserrate, often stopping by Simón Bolívar's home, the Quinta Bolívar, on the city's outskirts when they returned to town. For a representative of one of the world's largest oil companies, a diplomat such as Caffery was a particularly good friend to have. Ambassador Caffery and Ham remained close through the years, even as each married. Ambassador Caffery also developed a high esteem for Evelyn, following her artistic successes with great interest.

38
First-class method of travel to Barranca-Bermeja, Colombia.

Ham's verve and savoir faire made him one of the most popular bachelors in the capital. He spoke perfect Spanish, worked from 9 A.M. to 9 P.M. and enjoyed a very active social life. He frequently had seated dinners for twenty-four, serving the finest French wines and delicious food made from recipes acquired from his mother and the wives of friends. On these elegant occasions he decorated his home with large bronze vases of sunflowers, among the few blossoms that were rare in Colombia, where orchids were common. Evelyn perceptively notes that Ham combined the traits of both her parents—"he was ambitious and hard-working, yet fun-loving, gentle and patient."

## Lucky in Love

Although Evelyn was very intrigued with Ham, as he was with her, she continued to pursue her keen interest in art as their long-distance romance developed. After college she studied with Sally Farnham, training her eye and hand to create incredibly realistic portrait busts. Her bust of gallery owner Lionel Friend, one of her admirers from Palm Beach, is so accurate that the sun casts exactly the same shadows on the subject's face as on the finished artwork in a photograph taken in the garden of La Solana. Actually, Evelyn viewed Friend more as an adviser than a suitor. For instance, she consulted him on the important matter of whether she should follow her interest in art by developing her talents as an art historian or as an artist. His reaction was definite and unhesitating. "You should be an artist, by all means. You are much too talented to be happy devoting your life to studying others' work." Another career possibility which emerged after graduation from Vassar was the position of Executive Secretary of the Princeton Art Museum, offered to her by Frank Jewett Mather, the museum's director. Evelyn, who was very serious about making art a major component of her adult life, considered the offer for awhile but decided she would rather devote herself to painting.

In the early 1930s Evelyn also took life drawing classes at the Art Students League and the National Academy of Design. These sessions yielded a number of distinctive paintings of nude models, seen knitting or wearing such unusual headgear as turbans and picture hats.

Evelyn was planning an art tour of Italy with Maria Bizzoni, her Italian professor at Vassar, in the summer of 1934, when Ham suggested that they get married and go with each other instead. They had spent just eight days together before their marriage, but they had come to know each other's personalities, thoughts and dreams perhaps more intimately than other courting couples, because of the distance that separated them. The length of time required to learn about one another through letters and cables added immeasurably to the depth of their mutual attraction and understanding. Ham's enthusiastic descriptions of his fiancée led Alfonso López, then president-elect of Colombia, to make a special point of visiting Evelyn at the Waldorf when he was in New York in June 1934. The president's entourage included his wife, Maria, and aides, among them future Colombian president Lleras Carmargo. Maria López told Ham later that she found Evelyn *encantadora* and hoped she would be happy in Bogotá.

From the first days of their marriage they were drawn to each others' interests and eager to share their life's experiences. Because Ham had never been on one of the really big steamers, they took the *Conte de Savoia* to Europe for their honeymoon and returned on the *Île de France*. Traveling through Italy, Switzerland, France and England, they visited many of the important museums. Eager to learn as much as he could about Evelyn's favorite subject as quickly as possible, Ham read art books voraciously. Evelyn would later quiz him by covering up the museum labels and asking him to identify the artist.

After a three-month honeymoon, they returned to New York briefly before boarding the small Colombian liner *Pastores*—an experience as new to Evelyn as the *Conte de Savoia* was to Ham—and headed for Bogotá. Prior to their departure, Sheldon and Mary Whitehouse gave them a fabulous farewell luncheon in New York to which they invited close friends from Newport. One of the more entertaining luncheon guests was Alice Longworth, Theodore Roosevelt's witty

*34*
Lionel Friend in the garden of La Solana sitting next to his bust sculpted by Evelyn, ca. 1933.

*Evelyn particularly loved
the older sections of Bogotá,
filled with colonial churches
and homes.*

*Bogotá, Colombia, ca. 1937*

*35*
Opposite: Plaza and cathedral

*36*
Near right: Heart of the city

*37*
Far right: Quinta Bolívar on the outskirts of town

daughter. Sheldon, a former minister of the United States legation in Colombia, kindly provided Evelyn with some valuable insights about what to expect from life in Bogotá.

With a population of 375,000 in 1934, the Colombian capital really felt like a small town. "Everyone knew everyone, and everyone knew Ham," recalls Evelyn of her first impressions. She remembers an extremely cultured social scene—"anyone could get up and recite a sonnet." It was also a bastion of Gilded Age formality. Dinners among the elite group that ran the government, known socially as "La High," were lengthy black tie affairs; if women were present, they were white tie. After dinner, the ladies and gentlemen would take their coffee and cordials in separate salons, a feature of social etiquette that was extremely frustrating for Palma Guillén, minister to Colombia from the progressive Mexican government of Lázaro Cárdenas. She complained more than once to Evelyn and the other women that she was missing the crucial phase of the evening's gathering at which most of the important business and political transactions took place.

Colombia experienced great political stability in these years during which the democratically elected social elite ran the country. President and Mrs. Olaya Herrera were good friends of Ham and Evelyn's, as were later presidents. A further indication of Bogotá's formality, notwithstanding its small population, was Mrs. Olaya's request that Evelyn bring her back a full-length ermine coat when she next returned from New York—a favor Evelyn gladly provided.

One form of recreation they enjoyed was horseback riding. Among the Metzgers' riding companions were the children of the next president, Alfonso López. Ham frequently joined the young men of the family in daredevil antics on horseback—

riding without stirrups, jumping on and off the horse at full gallop. When Eva Borchard visited Bogotá later, President López and his wife invited the Metzgers and Mrs. Borchard to the palace for dinner. "How long will you be staying?" inquired the president.

"Two weeks," replied Eva.

"Why such a short visit after such a long trip?" he queried with some concern.

"I don't want to feel like a *suegra* (mother-in-law)," was Eva's explanation.

President and Mrs. López forthwith showed Eva a suite of rooms in the palace which they urged her to occupy for the winter, so she could visit Bogotá as long as she wanted without being a suegra.

Evelyn particularly loved the older sections of Bogotá, filled with colonial churches and homes. There she purchased early parchment-covered books, eventually amassing a collection of several hundred volumes—mostly on religion and law—which she later donated to the New York Public Library. Two of her favorites were a volume on early electrical experiments published in France in 1748, which depicts ornately garbed ladies and gentlemen serving as conductors for electrical current, and the first biography of Saint Ignatius Loyola, with engravings on each page illustrating the events of his life.

Travel provided additional recreation. Getting to know Colombia, Evelyn and Ham visited historic Popayán, Cali and Medellín, which enjoyed a more temperate climate than Bogotá and overflowed with wild orchids. They also frequented Cartagena, an extremely picturesque 16th-century city which was the headquarters of Captain Flanagan's Andian National Corporation.

In spite of having all the distractions of a young bride learning the ways of a new country, with Ham's constant support and encouragement, Evelyn continued to devote as much time as she could to painting. She produced a series of strong, insightful portraits of native residents of Bogotá. The care with which she treated their expressions and captured each detail of their wardrobe suggests that Evelyn shared something of her friend Malvina Hoffman's ethnographic interests. Evelyn also recorded the Colombian countryside in works such as her two distinct views of Cartagena—one realistic and one cubist—from the top of the Andian Building, prefiguring years of artistic experimentation. In addition to providing personal fulfillment, Evelyn's painting sometimes yielded other benefits in these years. For instance, after she painted the portrait of Ernesto Cano, owner of Bogotá's daily newspaper *El Espectador,* the paper's anti-American tone became much friendlier.

*Cartagena, Colombia, ca. 1937*

*38*
Top left: Andian Building, then the city's tallest structure.

*39*
Top right: Tropical streets

*40*
Bottom left: City's massive colonial walls, built 60 ft. high and 40 ft. thick for protection against pirates.

*41*
Bottom right: Tanker at deep-sea oil terminal of Andian Pipeline.

## Egypt, 1937

*42*
Evelyn (left), Ham and Eva on Avenue of the Sphinxes at Karnak.

*43*
Near right: Suez Canal just below Luxor

*44*
Far right: Portico of temple to Amon built by Amenhotep III at Luxor.

*45*
Vendors along Suez Canal

The Metzgers' travels also took them farther afield. In 1937 Evelyn and Ham joined Eva Borchard for a wonderful trip to Europe and the Middle East, visiting Austria, Italy, Greece, Turkey and Egypt. Hearing of their plans, Cornelia Weston, a cousin of photographer Edward Weston's and a Vassar friend of Evelyn's, insisted that Ham take along photographic equipment. She purchased a Rolleiflex for him, carefully compiled a list of additional paraphernalia needed and helped to secure all necessary accessories. The result of Cornelia's foresight, Ham's meticulous attention to detail and Evelyn's eye for framing subjects (she says she acted as art director for many of the photos) is the first in a series of stunning travel pictures produced by Ham and Evelyn from the late 1930s on. Crystal clear and filled with exquisite detail, they record Istanbul's domes and minarets, the monumental columns of Ba'albek balanced precariously over Evelyn's head, Egyptian traders selling their wares along the banks of the Suez Canal, dhows on the Nile and the regal splendor of the ceremonial passageway at Karnak. In addition to capturing exotic sites, the photographs also document the travelers' sophisticated style of dress. The elegant formality of their traveling attire—handsome, well-cut outfits with a hat or cap set at a jaunty angle—underscores the early date of the photographs, as does the absence of tourists.

In Egypt they took the customized Cook's "Night in the Desert" tour. On the appointed evening, Ham and Evelyn rode out seven miles into the desert on camels, but Eva, "an old lady of fifty-five," opted for the safety of a donkey cart. The ride seemed to go on forever, and Eva began to worry that she was being kidnapped.

"Where are you taking me?" she finally queried the Egyptian in charge of the expedition.

"Oh yes," came the reply.

"Is it going to take much longer to get there?" she asked, trying another approach.

"Oh yes," he said with an ingratiating smile.

"Are you kidnapping me?" she asked, with a distinct tone of fear creeping into her voice by this time.

"Oh yes," was all she heard.

Meanwhile, as the camel caravan proceeded, Evelyn noticed that Ham's camel kept sidling over and appeared to be eying her bare knee with evil intent. She nervously asked *their* guide whether Ham's camel was likely to bite her. "Oh yes" came the response.

*Egypt, 1937*

46
Evelyn (left) and
Eva at Thebes.

*The elegant formality of their traveling attire—handsome, well-cut outfits with a hat or cap set at a jaunty angle—underscores the early date of the photographs, as does the absence of tourists.*

## A Night in the Desert, 1937

*47*
Opposite: Evelyn poses on her camel while Eva stands resplendently in her fur-collared coat prior to setting out for the desert.

*48*
Above: Evelyn adeptly guides her camel past Eva's donkey cart.

*49*
Right top: Evening falls on desert caravan.

*50*
Right bottom: Dawn light strikes camp seven miles from the pyramids at Giza.

Arriving at their destination, the tour attendants set up camp—a tent for entertainment, one housing the kitchen and two bedroom and bathroom tents, along with tents for the staff—all very grand. Evelyn and Ham had been in the desert camp for what seemed like a good part of the night, and they, too, began to get worried about Eva. "A colossal red moon came up out of the desert," recalls Evelyn. "It appeared out of nowhere and gave way to a starlit sky." Finally, Eva arrived in the donkey cart with her sheik. Having reluctantly resigned herself to being kidnapped, she was more than a little relieved to be reunited with Evelyn and Ham.

After Egypt the threesome toured the European continent. They again experienced the special benefits accruing to art connoisseurs who were among the few visitors to the monuments and museums of the day. When they arrived at the Albertina in Vienna and expressed an interest in Old Master etchings and drawings, they were handed folios of original drawings and watercolors by Dürer, Brueghel and others to examine at their leisure. Evelyn remembers lingering over Dürer's famous *Praying Hands (Study for an Apostle)* and *The Hare*. "Today," she comments, "visitors are only allowed to view framed reproductions of these same works."

## Raising a Family

In 1938 Ham was transferred by Standard Oil to La Paz, Bolivia. Before leaving Bogotá he called on then President Eduardo Santos (the third president to govern since Evelyn's arrival) to bid him farewell and ask if there was anything he could do for him. Santos replied, "Leave me your cook, Josefina." She remained with the Santoses for years, even traveling with them to Paris.

As another foreign service friend, Freeman ("Doc") Matthews, was later to write in his memoirs, *Memories of a Passing Era,* Ham was such a capable representative for Standard Oil that he became "their Latin American trouble shooter." Many Latin governments, starting with Bolivia in 1938, were intent on expropriating, or threatening to expropriate, the property of American businesses. One of Ham's principal jobs was to dissuade governments from taking such action against Jersey Standard or, failing that, to negotiate compensation for expropriated properties. Assigned to the petroleum hot spots of Latin America, the Metzgers began a peripatetic existence, living in four different countries and ten different homes over the next six years.

*Views of Europe and the Middle East, 1937*

*51*
Right: Eva (front) and friend touring Pompeii, Italy, in sedan chairs.

*52*
Middle: Temple of Jupiter, Ba'albek, Lebanon.

*53*
Bruges, Belgium

54
Evelyn on the Acropolis,
Athens, Greece.

55
Evelyn holding up column
of Temple of Bacchus at
Ba'albek, Lebanon.

Evelyn describes La Paz as "a colorful place at the top of the world." At 13,000 ft., it shares the honors with Lhasa, Tibet, as the highest capital in the world. Consequently, the city has a climate in which one can be freezing in the shade and boiling in the sun. Whether due to fashion or comfort, local inhabitants favor the layered look. The Indian women of Bolivia wear six or seven woolen skirts of different colors and the men have caps and cloaks of the brightest hues.

Once settled in La Paz, the Metzgers soon made friends with the country's tin mine owners, such as Carlos Victor Aramayo, who at one time held the distinction of being the world's wealthiest man. Having grown up in England, Aramayo was further distinguished by being a golfing companion of Edward VIII when he was Prince of Wales. Another of the tin barons, Mauricio Hochschild, bet $50,000 in the late '30s that a United States of Europe would be formed within ten years. Needless to say, he had to pay up, although the emergence of the European Economic Community seems, belatedly, to have established his credentials as a visionary. The Metzgers' friendship with Hochschild continued after they left Bolivia. In later years he had the endearing habit of bringing them a pound of caviar whenever he dined at their New York apartment.

When they weren't exploring La Paz, the Metzgers would visit Lake Titicaca, which is shared by Bolivia and Peru. Sitting at 12,500 ft., nestled among the peaks of the Andes, it is the largest freshwater lake in South America and the highest in the world. The Metzgers often crossed the lake into Peru on the 1,000-ton *Ollanta,* a steam-powered boat that had been carried up to Titicaca in pieces, on horses and mules. They spent idyllic hours picnicking among the pre-Columbian ruins at Tiahuanaco in Bolivia and photographing the Indian market at Pisac in Peru. Some of these trips resulted in paintings. The distinctiveness of Evelyn's artistic perception is evident in her original treatments of Andean natives. One of the most unusual represents an Indian mother breast-feeding her child. Painted cubist style in shades of black, green and ocher, the work presents the planes and angles of the mother's black hat and body—shapes that shift and rejoin before the viewer's eye to create varying patterns that convey a veiled image of tenderness.

Evelyn learned that she was pregnant shortly after they arrived in Bolivia. Since the local clinic had no running water or other such amenities, Ham felt that she should return to New York by steamer to have the child. Their route to the coast took them across Lake Titicaca to Puno. There they boarded the highest railroad train in the world, running to 17,000 ft. between Arequipa and Cuzco. "Cuzco was at its best then, before the earthquake of 1950," says Evelyn. "It's a marvelous colonial city with pre-Inca structures of colossal stones fitted together with

*Bolivia, 1938*

*56*
The Metzgers' Bolivian cook and her baby.

*57*
Right: Lake Titicaca

*58*
Below: Laja's renowned plaza, near Lake Titicaca.

*Peru, 1938*

*59*
Pre-Inca architecture, Cuzco

*"Cuzco was at its best then, before the earthquake of 1950. It's a marvelous colonial city with pre-Inca structures of colossal stones."*

*60*
Near right: Plaza de Armas, Cuzco

*61*
Far right: Peruvian mother and child near Cuzco.

*62*
Right below: Templo del Sol, Cuzco

*63*
Sacsahuaman

no mortar. Only stone implements were utilized. It was a thrill to be there." After a few days in Cuzco, Ham took her to Callao, Lima's port, and put her on the ship for home. The first port of call was Buenaventura, Colombia, the wettest place on earth with 365 inches of rain per year. From there the ship went through the Panama Canal and on to New York.

The Metzgers' first child, James, was born in New York in August 1938. By luck, Ham was on hand for the birth of his son, but returned to South America shortly thereafter. In these months he was shuttling among Buenos Aires, Montevideo and New York. Evelyn remained with her baby for eight months, until the separation from Ham became too much to bear. Then, leaving Jim at the Waldorf with Eva and her second husband in attendance, aided by a full-time nurse, Evelyn went to Buenos Aires (known to residents as B.A.) to be with Ham for three months. They took up residence in the beautiful Plaza Hotel in the heart of the city.

During this stay in B.A. Evelyn renewed her acquaintance with the flamboyant Suzette Telenga Ellinger. Passionate and self-directed, Suzie was the type of woman who was spoken about as much for her exploits as for her accomplishments, perhaps more. Her first marriage to Ham's friend, the diplomat Alan Dawson, had ended in divorce, followed by his suicide. After her divorce from Enrique Ellinger, Suzie was destined to marry twice again. She also became the author of numerous romance novels and screenplays.

In her characteristically direct manner, she asked Evelyn, "What do you do?" when they met again in B.A. "I paint," came the reply. Ever the activist in her own life, Suzie thought that, if Evelyn was a painter, she should paint. If a subject was lacking, she could provide it. She forthwith helped Evelyn secure paints, canvas and other necessities, sent her chauffeur to collect her at the Plaza and began to pose for portraits. Through the years Suzie was to be Evelyn's most willing and patient model, sitting for dozens of likenesses.

Returning to New York in the fall of 1939, Evelyn, Ham and little Jimmy moved into an apartment in the Borchards' building at 815 Park Avenue, furnished it and settled down to life in the city. One day Ham came home and said,

"How would you like to go to Rio?"

"I'd like to stay right here," came the reply, "I just fixed up the apartment."

"Be reasonable, Evelyn," said Ham, and Rio it was.

64

A park in the center of Buenos Aires backed by the Cavanagh Building (left) and Plaza Hotel.

In 1940 the city had the same breathtaking beauty it has today, without the crowds. "I was mad for Brazil. I loved it!" recalls Evelyn. She was delighted that her Colombian friend Maria Helena Samper de Donnelly and her husband, the American diplomat Walter Donnelly, were in Rio at this time. Through Maria Helena's efforts, the counselor of the United States embassy made available his home, fully furnished, to the Metzgers during his six-months' home leave.

Evelyn was pregnant again and, with Ham out of the country, Maria Helena made arrangements for the birth at the Maternidade Arnaldo de Moraes in Copacabana. The Metzgers' second son, Edward, was born in July 1940. When he was just three months old the family, ever on the move, left Rio for Santiago, Chile. Evelyn still is amused that the United States consulate found it necessary to fill in the marital status on her baby son's first passport as "unmarried."

Evelyn's principal recollections of her thirteen months in Chile revolve around its inhabitants. "They are very loyal people. They never forget you, and are extremely hospitable. From the day we arrived we were invited to someone's home. They spared no effort, gave parties often and were very informal and straightforward." Of course, she also remembers the spectacular Chilean countryside, the lake section and the snow-covered volcanoes. "They are incredibly beautiful."

*The Lake District of Southern Chile, ca. 1940*

*65*
Top: Calbuco volcano

*66*
Middle: Osorno volcano

*67*
Bottom: Puerto Montt

All too soon, it was time to leave Chile and return to Rio. There, in May 1944, their daughter was born. Evelyn and Ham named her for her maternal grandmother. It's no wonder that all the members of her family have loved to take trips says Evelyn. "I traveled before I was born and so did all my children." When Ham was transferred to Buenos Aires shortly after Eva's birth, they found that wartime agreements reached between the governments of Getúlio Vargas and Franklin Roosevelt prohibited Brazilian women from flying out of the country. Luckily, their friend Jefferson Caffery was still ambassador to Brazil. In his final months there, Ambassador Caffery worked successfully to secure a decree permitting "Brazilian women" small enough to sit on their mothers' laps to fly, and the Metzgers were free to take Eva with them to their new home.

## Argentina—A Time of Contrasts

In 1944 Ham Metzger became president and general manager of Standard Oil's Argentine affiliates (as well as affiliates in Uruguay and Paraguay)—one of Standard's pivotal posts in South America. After many years on the move, the family was to have an opportunity to settle down in Buenos Aires, living there for the next ten years.

In the mid-1940s Argentina was at a crossroads. The country had maintained its neutrality during World War II, despite strong Axis sympathies. With a growing market for its natural resources and cattle, Argentina enjoyed a prosperity in the late 1930s and early 1940s virtually unsurpassed in its history. But the seeds of repression and economic stagnation had been planted in 1943, when a popular army colonel named Juan Perón helped to overthrow the government of President Ramón Castillo. Only one of numerous military men vying to control the Argentine government at this point, Perón had the insight to enhance his political profile by allying himself with the country's labor unions. He became Secretary of Labor and Welfare at the end of 1943, a position he retained when he ascended to the vice presidency in July 1944. Like other authoritarian world leaders of the period, he could attract and mobilize masses of people through a combination of a mesmerizing charisma and instinctual manipulation of the media. It was a general strike and massive union demonstration in October 1945 which catapulted Perón, "Argentina's Number One Worker," to uncontested national prominence. Early in 1946 the same unions swept him to victory as Argentina's twenty-ninth president. The United States ambassador to Argentina, Spruille Braden, was adamantly opposed to Perón. Yet Braden's anti-Perón tirades were

68
Top: Evelyn and Maria Helena Donnelly entertaining their sons Jim and George on Copacabana Beach, Rio, ca. 1940.

69
Middle: Ed's first birthday

70
Bottom: Ed at nine months

71
Top: Ed and Jim at Rio beach, 1944.

72
Middle: Jim's third birthday

73
Bottom: Jim and Eva, ca. 1946

74
Evelyn and Eva in their Buenos Aires pool.

75
Below: Bedtime in Buenos Aires

76
Among the children on hand to celebrate Eva's birthday at her home in Buenos Aires, were Martha Argerich (lower right), who has since become an internationally renowned concert pianist, Eva (center) and Ed and Jim (top row).

*"Martha transfixed us with her keyboard brilliance. She was otherwise a typical child, frolicking in our pool."*

77
Ed (left), Eva and Jim in Buenos Aires.

*Plate 4*
*Llao Llao Trees*
1949
Oil on canvas
27 × 19½ in.

*78*
Paul Cezanne
*Chestnut Trees at Jas de Bouffan*
1885–87
Collection The Minneapolis
Institute of Arts

widely credited (or blamed) for getting him elected. "Braden o Perón" was the slogan throughout the capital as election day approached. Ambassador Braden and his wife Maria, who were transferred shortly after Perón's election, were close friends of the Metzgers while they were in Argentina, and remained so later.

The Metzgers' years in B.A. were filled with the pleasures of family, friends and life in a glamorous, cosmopolitan world capital—"a place of sumptuous boulevards, opera, and women of high fashion, a distillation of France and Italy and continental elegance," in Robert Crassweller's words. James, Edward and Eva learned to swim and dive in the family swimming pool, made friends among the children of their Argentine, American and British neighbors, were rambunctious at every opportunity and, in general, behaved like healthy youngsters. The pleasure they took in the culture surrounding them is clearly evident in a photo of the three of them decked out in full cowboy regalia, with a few gaucho touches thrown in.

With her children no longer requiring constant attention, Evelyn was able to devote more time to her art, painting numerous portraits of friends and acquaintances. She also experimented with a new theme—street scenes of B.A. painted in a dusky palate of grayed pinks, blues and greens which capture the romantic, Old

*Plate 5*
Three Carriages, Belgrano R
1949
Oil on canvas
19½ × 23¼ in.

*Plate 6*
*San Isidro Church, Buenos Aires*
1950
Oil on panel
23 × 31 in.

World allure of the city. She produced dozens of portraits of her children, awake and asleep, individually and collectively. Daughter Eva, who frequently was the subject of her mother's art in these years, says that being a model wasn't entirely unbearable, because her mother worked so quickly. Evelyn often caught her on the fly, reading a book or otherwise entertaining herself. "She'd just say, 'hold it, hold it,' and get to work." Another subject was her children's friends. Evelyn recalls particularly a double portrait she made of Jim and his boyhood friend and classmate John Reed playing chess. John lived across the street, so the two boys saw each other a lot. Jim would go to John's house for Coca-Colas (which were banned at the Metzgers'); John, in turn, would visit to swim in the Metzgers' pool. The boys remained good friends, even pursuing similar careers. Jim's years in international finance led to his appointment as director in charge of the United States branch of Banco de Bilbão in New York; John rose through the ranks to become CEO of Citicorp.

Edward remembers that on most days when he arrived home from school his mother was in the studio. "Painting was nearly a compulsion for her—she has a great love for it," he explains. "But she never allowed it to distract her from other activities. I never had the impression that I could not ask for her attention, nor that my father did not receive attention, because of her art." William and Florence Leonhart, United States diplomatic corps friends in B.A. at the time, also were impressed with Evelyn's devotion to her muse. "What I find extraordinary about Evelyn," remarks William Leonhart, "is her professional attitude toward her art, almost ruthless. In Buenos Aires she painted for hours every day and studied under accomplished and severe maestros."

In 1950 all of this hard work paid off, and Evelyn had her first solo exhibition at the prestigious Galería Muller on Calle Florida. The show was a great success, generating many positive press reviews. In conversation Evelyn tends to deflect attention politely away from herself and her achievements. Thus, her comments about her first exhibition are reserved for the huge quantity of flowers sent by her friends to celebrate the event, making it look "more like a flower show than an art show." The gratification and pride she felt on this occasion is subtly conveyed in her tone of voice.

B.A. offered an array of activities for Evelyn and Ham. They enjoyed walking along Florida in the center of town, a street filled with beautiful shops and closed to automobiles, which made it the perfect location for a relaxing promenade while admiring high-fashion *porteños* (as Buenos Aires residents refer to themselves). They liked to explore the area around their home in the B.A. suburb of Vicente Lopez, located about fifteen minutes from town. "We would walk down

*Plate 7*
Ham Metzger
1946
Oil on canvas
18 × 14 in.
Collection Edna M. Kane

*Plate 8*
*Suzie in Profile*
1946
Oil on canvas
20 × 16 in.

*Plate 9*
Suzie, Mirror and Lamp
1947
Oil on canvas
30½ × 23 in.

79
Left: Calle Florida, the main shopping street of Buenos Aires. UPI/Bettmann Newsphotos.

80
Right: The Jockey Club's San Isidro racetrack, Buenos Aires.

from our house just a few blocks to the river, and stroll along there," Evelyn says. "People were having a wonderful time. They were barbecuing, playing music and dancing the tango—activities that contrasted dramatically with their usual restraint." Ham and Evelyn had a box at the Opera Colón, seeing fourteen performances each season which featured the world's preeminent stars. Evelyn attributes her later lack of enthusiasm for opera to the experience of spending months of every year sitting in their box until 1 A.M. Ham, always more musically inclined and a dedicated night owl to boot, enjoyed these evenings immensely.

Ham was honored to be made a member of the exclusive Jockey Club, a membership he maintained even after leaving Argentina. The Metzgers attended races at the club's tracks in Belgrano and San Isidro. There were also two splendid eighteen-hole golf courses on which Evelyn won many of her cups and trophies. One year she played so much golf that she hardly had time to paint. "That's when I decided to give up competitive golf," she says.

"They were a devoted and extraordinarily handsome pair," says Ambassador Leonhart recollecting the Metzgers in B.A. "Ham was always dressed in Savile Row style, refined and very debonair—the most unlikely oil field roustabout one could imagine. Evelyn was a woman of style, grace, intellectual vitality and devotion to her husband. Ham admired her, and she Ham. They were together constantly."

81
Ham (right) in animated conversation with Emile E. Soubry, Jersey Standard's vice president and director, in May 1947.

82
Evelyn and Robert T. Haslam, executive vice president and director of Jersey Standard, show off the results of a fishing foray in Bariloche, Argentina.

> *"Ham was always dressed in Savile Row style, refined and very debonair—the most unlikely oil field roustabout one could imagine."*

---

Evelyn and Ham had countless friends in Argentina. Among them was the Grand Duchess Marie, granddaughter of Tsar Alexander II and cousin of Russia's last Tsar, Nicholas II. She and Evelyn often lunched together and discussed their painting. The Grand Duchess made paintings of home interiors, but did not give great importance to her work. She was a writer and a raconteur who had many stories to tell about her first days in New York. Once, while living in a borrowed apartment, having recently escaped from Russia, where her family had been executed, the doorbell rang. She opened the door and confronted a man who said, "I am the exterminator." At these words she felt the blood drain from her face and a sensation that she was about to faint come over her. The man then added, "I'm here to spray the kitchen." As the Grand Duchess explains in her two-volume autobiography, she grew up with a strong sense of impending doom, due to the danger of assassination by Russian extremists, who continuously threatened the lives of close family members.

Among their dearest friends were Alicia and Goar Mestre. Alicia was a beautiful, intelligent Argentine woman who had taught Spanish at Vassar. Goar, who had been born in Cuba and graduated from Yale, was a successful real estate developer and the owner of numerous television and radio stations. After Fidel Castro confiscated one of the gems of the family's telecommunications network—Radio Centro in Havana—Goar established Channel 13 in Buenos Aires. Politics being what they were in Latin America in the 1960s and 1970s, this investment proved not to be secure either. During his second tenure in office Juan Perón confiscated Channel 13, although restitution eventually was made to the Mestre family.

Lucrecia Moyano de Muñiz, whose family developed the famed Argentine beach resort Mar del Plata, rented them her home when they arrived in town, and became a friend. When they outgrew Lucrecia's charming pink house in Acassuso and moved to Vicente Lopez, they were just three blocks from Suzie. "She had a magnificent residence with a tennis court and a swimming pool—our pool was a baby compared to hers," Evelyn remembers. "We spent a lot of time with her, in her lovely house. It later became an embassy." They knew Hiram Bingham IV, son of the senator and archaeologist of Machu Picchu fame. One story about them that Evelyn recalls is that friends telephoned Mrs. Bingham one day to alert her to the fact that some of their twelve children were playing on the roof of their house. She forthwith called out to them, "Children, you must come in now, you're worrying the neighbors." Evelyn adds, "They left two of their children at our house after a birthday party—they forgot about them completely." Laura Riedel, "Mrs. General Motors" was a good friend and model of Evelyn's, as was Jean Boley, a published author and *New Yorker* correspondent, who was married to the head of International Harvester.

They also counted Francisco and Mercedes Cambó, and their only child, Helena, as good friends. Formerly Spain's progressive Minister of Public Works, Francisco Cambó had been among the first political leaders exiled by Francisco Franco in 1937. Evelyn got to know this avid art collector when he was critically ill and vacationing in Córdoba, Argentina. One of his few remaining pleasures at this point in his life was to talk with Evelyn at length about shared interests and mutual acquaintances in the art world. When the Franco years finally ended and Juan Carlos took charge, the Cambós' impressive collection of Old Masters went to the Prado and to Barcelona museums, a fitting tribute to Don Francisco's deep and abiding love for his country. This gift did not proceed without a hitch, however, since Perón refused at first to allow many of the collection's masterpieces, which were then in Argentina, to leave the country. It took the efforts of Helena and her husband, a leading member of Opus Dei, to convince Perón to release the works to Spain.

83
Suzie in the garden of her palatial Buenos Aires home, 1945.

84

Ignacio Zuloaga, *Retrato de Cambó* (Francisco Cambó). Collection Guardans-Cambó, Barcelona.

*Evelyn recalls visiting Zuloaga's home in San Sebastián during her trip to Spain in 1932. While conversing on this occasion, the artist told her that he had been instrumental in reestablishing El Greco's reputation. Difficult as it may be to imagine today, El Greco had been unappreciated for centuries.*

Among their other Argentine friends were several who owned stunning *estancias* or country houses, often situated on vast tracts of land. Estancias to which Evelyn and Ham were frequently invited included the Pereira Iraola estate near Buenos Aires and Enrique Larreta's magnificent Acelain in Tandil.

Beneath the surface of pleasant daily pursuits in B.A. ran the steady current of Peronism. As representatives of a major U.S. company, Evelyn and Ham met the president and his wife Evita on several occasions. Once when Perón dined at their home he wagered a custom-tailored suit of clothes with one guest that the United States would be at war with Russia within the year; Perón paid up. Evelyn remembers an event at the United States embassy during which Evita remarked to her, "I have two things to be proud of in life—the love of the people and the hatred of the *oligarquía*." When Evelyn demurred Evita replied, "Señora, if it were not true, I would have to invent it." Evita was quite conscious of the political value of the contempt expressed for her by the country's upper classes.

"The cult of Evita was taking over Argentina," recalls Evelyn. "They were making a saint out of her. You would see people praying to her portrait." The Jockey

Club became a target for demonstrations against the social elite. This luxurious and historic landmark on Calle Florida, famed for its excellent art collection and superb library, was burned to the ground by militant Peronists in 1953. The police and fire department had instructions not to intervene. Carlos Aloé, one of Perón's principal aids, summarized the political realities of the day by the mid-1950s. "In the Argentine government, there is no one, neither governors, nor deputies, nor judges, no one; there is one government only, and it is Perón."

In 1954 Ham and Evelyn decided they should return to New York to live, so that their children could continue their education in the United States. With months to make arrangements, Evelyn planned a dream trip—a leisurely world voyage on the MV *Boissevain* of the Dutch Royal Inter-Ocean Line. They started the trip on an entertaining note, running into movie stars Fred MacMurray, June Haver, Marlon Brando, Walter Pidgeon and Irene Dunne at various restaurants and hotel lobbies in São Paulo. The celebrities were in town to promote the newly established São Paulo Film Festival. In South Africa they visited a Ndebele village, where they were impressed by the vibrant geometric patterns that the women of the tribe paint on the interiors and exteriors of their houses and the thick beaded rings the women wear around their necks, which Evelyn compared to scooter tires. At the Johannesburg gold mines they were entertained by the miners' dancing and intrigued to notice that the policemen—on hand to keep order—could

85
Juan Perón waving to a rally of the Argentine Confederation of Labor Unions in June 1955. UPI/Bettmann Newsphotos.

86
Eva Duarte Perón, 1944.
UPI/Bettmann Newsphotos.

not resist joining in when members of their own tribe performed. On the Asian continent, colorful junks and sampans dotted Hong Kong's harbor, while rickshas filled the streets. Just two skyscrapers were prominent on the city's horizon—the headquarters of the Hongkong and Shanghai Bank, dating to the mid-1930s, and the recently completed Bank of China—leaving the majesty of the surrounding hills and peaks clearly visible. These sites and many others filled their eyes and memories as each day progressed. When the family arrived in San Francisco three months after leaving B.A., two Standard Oil limousines were waiting to take them to their hotel and company representatives were on hand to help organize the remainder of their journey home.

## Based in New York

Once back in New York, Evelyn's immediate concern was her mother, who was seriously ill. In the months that remained of her life the two women spent days together, bridging, in some measure, the twenty years of separation. Eva died in 1955.

*87*
Evelyn and her children.

Getting on with her life after her mother's death meant, for Evelyn, getting on with her art, and the late 1950s were the beginning of her most productive period, which continues to this day. Ham's career also proceeded apace. In 1954 he became vice-president and director of Creole Petroleum, in charge of the New York office. "Jersey's major affiliate in Venezuela, Creole Petroleum Corporation, was the principal source of crude production and the company's single most profitable operation until the mid-1960s," notes Bennett H. Wall in his Standard Oil history regarding the period of Ham's association with Creole.

The Metzgers continued to take interesting trips. In 1958 Evelyn took her daughter, Eva, to Europe. She hoped to share some of the experiences with her own daughter that she and her mother had so enjoyed on their 1932 tour. When they arrived at the Louvre, Evelyn's favorite rooms near the Rubens galleries were closed. After making a fervent request in French to the administrative officials, Evelyn had the satisfaction of having the rooms opened briefly so Eva could view these treasured objects. Having seen Evelyn's favorite pieces, the two then went on an extensive tour of the Louvre. Asked to recall this experience, Eva says that, at the time, she found it a little difficult. "I always think about it in astrological terms. I'm a Taurus, and, so I've been told, Tauruses go vertically into one thing. Mother's a Gemini—they go horizontally. At the Louvre, she wanted to see everything and I wanted to concentrate in one area, spend some time and really get into it. So I felt a little frustrated, because we would rip through, having to see everything. In retrospect, however, I appreciate the experience because I did get a lot of art education from her, and looked at things I never would have seen on my own."

Reprising a portion of her graduation tour of 1932, Evelyn took Eva to visit the prehistoric caves of the Dordogne, including Lascaux and many others. Eva's main impression of her mother in this context is her fearlessness and stamina when called upon to crawl on her stomach through the narrow openings into some caves, or scale dizzying walkways made slippery by dripping stalactites. "In contrast to her quiet demeanor at home, her daring bordered on recklessness when we traveled," notes Eva. "She didn't want to miss anything. Later she traveled with Eddie to places I would never go. She didn't mind hardship one bit."

Ham joined them in Rome in time to accompany Jefferson Caffery, who was Honorary Papal Chamberlain, to the Ninth Mass preceding the selection of the next pope. This brought back memories of the private audience she and Ham had with Pope Pius XI on their honeymoon in 1934. In Caffery's suite at the Grand Hotel they met his good friend Clare Boothe Luce, former United States ambassador to Italy, who was "colorful, attractive and striking in a grey dress," recounts Evelyn. Luce predicted that Angelo Giuseppe Roncalli, who had been Papal Nuncio in Paris when Caffery was ambassador there, would be pope. Luce declared jovially, with a wink at Caffery, that the basis for her prediction was the fact that she had once seen the Nuncio kiss the ambassador on both cheeks. "Anyone who would kiss Jefferson Caffery deserves to be pope." Luce's intuition was correct and Roncalli was elected John XXIII.

88
The Metzgers dressed in traditional attire during their 1954 visit to Japan.

A year later Evelyn launched a tanker—the *Esso Amuay*—in Japan. The Metzgers took Eva out of the Brearley School for this event. At the insistence of her Japanese hosts, Evelyn repeatedly practiced severing the cord with a small axe, so that no problem would arise during the actual launching. With typical thoroughness, the shipbuilders also had a man hidden beneath the launching platform ready to sever the cord at the appointed time if all else failed. At 45,000 tons, the tanker looked huge, although it was "quite small," says Evelyn, "by today's standards." Happily, everything went off without a hitch. "Once the cord was cut, a Rube Goldberg-series of actions caused the tanker to slide gracefully into the water. I don't know how many hundreds of doves were let loose or thousands of schoolchildren were on hand cheering," recalls Evelyn. "I even said a few words of Japanese successfully."

The Japanese shipbuilders provided every amenity during their stay, including an extensive tour of Japan by private train for the family and their entourage of hosts. William and Florence (Pidge) Leonhart, the Metzgers' good friends from B.A. then stationed in Tokyo, were able to join the group and help introduce the splendors of Japan to Evelyn, Ham and Eva. "What stands out about Evelyn on this trip was her eagerness to absorb the culture," says William Leonhart. "She asked lots of questions and was unusually observant." At one of the country's most celebrated papermaking and printing companies, Evelyn and Ham decided to have their Christmas cards printed using a photograph from the trip, which would allow them to share their experience with more of their friends. They tended to order cards early because their list numbered eight hundred names in these years, and they had to begin addressing them in September. After seeing the beauties of Japan, the family took a picturesque route home—visiting Thailand, Cambodia and Turkey.

In 1962 Ham retired from Standard Oil to join the Houston-based international engineering firm of Brown & Root. Writing to Jefferson Caffery (the "Chief") in January 1963, he explained, "This new employment offers a pleasant transition between the 24-hour day with Jersey and 100% idleness." He then added with a touch of levity, "Evelyn has become so busy with shows . . . that she, too, is pleased I have an office to go to and can thereby keep out of her way." Writing from the Grand Hotel, Caffery sent Evelyn his warm congratulations and encouraged the Metzgers to visit him and Gertrude in Rome following the opening of her show in Paris in June 1963.

Evelyn, in fact, was extremely busy with exhibitions during these years. From 1963 to 1969 she had fifteen solo shows which embellished the walls of Vassar, museums up and down the East Coast, the Galerie Bellechasse in Paris and the

Mexican-American Cultural Institute in Mexico City. A highlight of the Mexico City exhibition was having George and Alice Brown fly the Brown & Root company jet down from Houston for the opening, bringing numerous mutual friends. Renowned philanthropists, the Browns endowed the Brown Pavilion of the Museum of Fine Arts, Houston, which Alice showed Evelyn before its formal opening; the annex was larger than the rest of the museum. They later donated the George R. Brown Engineering Building at Rice University. George and his brother Herman Brown were Lyndon Johnson's chief financial backers, and indispensable to his political success.

Another reason Ham gave Caffery for joining Brown & Root was his hope to take three months off each year to travel. While Ham never took as much leave as he intended, the 1960s were filled with trips. The Metzgers traveled through Europe in 1963; to India, the Middle East and northern Africa in 1965; to the Canaries and Costa del Sol in 1966; to Mexico in 1967; throughout the Pacific and Southeast Asia in 1968; and to East Africa—Ethiopia (including the ancient capital of Axum and Lalibela with its 12th-century churches dug down into the live red rock), Kenya, Zambia and Tanzania, including Zanzibar—in 1969. Evelyn has a map in her guest room in which she has inserted color-coded pins representing trips by various family members; some of the continents look like miniature porcupines.

At a dinner Ham and Evelyn gave in 1965 in their home for Roger Blough, CEO of the United States Steel Corporation, and M. J. Rathbone, CEO of Jersey Standard, Ham mentioned his forthcoming trip to North Africa.

"You're not going to Libya?" inquired Rathbone.

"They don't welcome tourists," said Ham.

"You wouldn't be tourists," replied Rathbone, "you'd be guests of Standard Oil."

This conversation led to a week's stay at the company's fully staffed guest house in Libya. Representatives of the company flew the Metzgers to Cyrene, where they were enthralled by the extraordinary Greek ruins dating from the 7th century B.C. of the once-thriving metropolis in which a million people had lived. "We were truly privileged travelers because of Standard Oil," says Evelyn recalling this and other trips. "We were never alone, never tourists, always had a family to take care of us, to show us the best of everything."

Evelyn remembers a fascinating evening with Noël Coward from a trip to the South Pacific three years later. At dinner in Bora Bora, Evelyn was seated opposite

89
The launching of the *Esso Amuay*, 1959.

---

*"Once the cord was cut, a Rube Goldberg-series of actions caused the tanker to slide gracefully into the water. I don't know how many hundreds of doves were let loose or thousands of schoolchildren were on hand cheering."*

*90*
Evelyn ceremonially receives the axe she used in launching the *Esso Amuay*, 1959.

*91*
A bevy of young geishas formed part of the entourage of hosts for the Metzgers and their friends at the 1959 launching of the *Esso Amuay*.

*92*
Novelty photograph of Evelyn as flamenco dancer and Ed as guitarist taken on a trip to Spain, ca. 1958.

*93*
Manolo Barturen, one of Jim's early employers, introducing Eva to the Duchess of Badajos, sister of Juan Carlos, king of Spain, ca. 1962.

Sir Noël. He was most entertaining, telling of his visits to Elizabeth II at Buckingham Palace—how friendly and cordial she was. Evelyn mentioned swimming out to a particular coral head several times and always seeing the same brightly colored little fish with their distinctive blue, yellow and red markings. Sir Noël rejoined, "Of course you saw the same fish. You must read Ardrey's *The Territorial Imperative,* which explains that all creatures have their own 'territory' or living space."

Evelyn and Ham now had much more time to spend with their New York friends, such as Lydia Winston Malbin. An insightful and challenging art collector, Lydia had the courage to amass a unique collection of Futurist and Constructivist art beginning in the 1950s. Since Futurism was a relatively short-lived movement which never gained prominence in commercial galleries, Lydia was forced to locate works on her own, purchasing nearly her entire collection directly from the artists or their families. Although they were opposites in temperament, personality and viewpoint, Evelyn had great fondness for Lydia and credits her with originality and prescience. After her death Sotheby's sold ninety items from her collection in May 1990 for $70 million.

More pins were added to the travel map in the early 1970s. Among their destinations were Portugal, Yugoslavia, Russia and the Mayan centers in Guatemala and Honduras. The Yugoslavia trip stands out because they had an opportunity to stay with their friends Ambassador and Mrs. Leonhart at the American embassy residence in Belgrade. Bill and Pidge were wonderful hosts who carefully arranged their visit through the country. Some of the notable sites Ham and Evelyn saw were the old town of Mostar with its colorfully painted houses and unique bridge. Evelyn also was taken with the frescoed façades of some of the medieval Yugoslavian churches.

Evelyn's last trip with Ham was to Turkey in 1973—a trip inspired by Turgut Menemencioglu, Turkish ambassador to the United Nations, who had told them

---

*"We were truly privileged travelers because of Standard Oil. We were never alone, never tourists, always had a family to take care of us, to show us the best of everything."*

94
Evelyn Metzger, ca. 1962

95
Ham Metzger, ca. 1962.
Photograph by Fabian Bachrach.

not to miss the archaeological wonders of his country. With guidance from their Turkish friends, the Dirvanas, Ham and Evelyn took an extended motor tour, visiting thirteen archaeological sites, including several pre-Hittite ruins.

Ham died of cancer in 1974. One of the tributes to her husband that Evelyn values most came several years later from Wallace E. Pratt, the legendary geologist who became vice-president and director of Jersey Standard. Bennett Wall calls Pratt "an oil giant—an all time great who enjoyed to the end of his long life an almost total recall of events—a shrewd judge of men." Writing to his friend Mrs. Newton Bevin, Pratt expounded on his impressions of Ham Metzger. "He was perhaps the most erudite man I ever knew. . . . He was the most courteous of men." Pratt then added that, in his view, Ham had all of the qualities necessary to have become Chief Executive Officer of Exxon Corporation (the company Jersey Standard had evolved into by then).

Having been such a close team for forty years, it took some time for Evelyn to regain her former joie de vivre after Ham's death. She eventually turned to her friends for solace and support. One friend who helped her through this period was Lucrecia Moyano. Now over ninety, Lucrecia is a successful glass and textile artist who was recently honored with an exhibition at the Museo de Artes Decorativos in B.A. She and Evelyn still maintain a friendship begun fifty years ago in Argentina. Lucrecia visits Evelyn yearly in New York and Quogue, L.I., health permitting. Evelyn began seeing more of Lydia Winston in New York, often exchanging weekly visits. At Lydia's dinners and receptions she frequently encountered such art world luminaries as Anne d'Harnoncourt, the Director of the Philadelphia Museum of Art whom Evelyn has known since she was a student at the Brearley School, and Sir John Pope-Hennessy, the world-renowned Consultative Director of European Paintings at the Metropolitan Museum of Art and authority on the Italian Renaissance.

Since Ham's death, Evelyn has spent a good deal of time with her children. Asked to describe her mother's most salient characteristics, which have become more evident in recent years, Eva cites humor. One episode she recalls with particular vividness took place at Evelyn's summer home in Quogue. Distraught about the failure of her first marriage and her pending divorce, Eva had gone to Long Island to try to relax. Instead, she spent days, seemingly, telling her mother that she didn't know what she was going to do and that she would never fall in love again. After listening sympathetically for hours, Evelyn responded, "But, darling, it's only been two weeks!" This upbeat remark somehow cast an entirely different light on the situation. They laughed together about Eva's melodramatic portrayal of her prospects, and went on from there. In fairly short order Eva was happily remarried.

It was at this time that Evelyn began work on the series that have occupied her attention for the last two decades—the unlimited moods and vistas of Central Park, portraits of her children and grandchildren documenting the stages of their lives and scenes of travel. Some of Evelyn's recent works record the journey to China which she and Edward took in 1979, when it was first possible to visit there again. The novelty of the experience for both cultures is revealed by Evelyn's recollections of the trip. When the group traveled to outlying villages the townspeople would stop everything and stare at them. "We were terribly curious about them, but they were more curious about us," she notes. They visited Tatung, a colorful ancient city in northern Shansi province near Beijing. About ten miles outside of town are the fabled cave temples of Yün-kang, an early religious sanctuary of more than forty grottoes filled with Buddhas and other religious statuary. Evelyn was intrigued by the blending of artistic and spiritual influences in these works. Another highlight of the trip was their visit to the archaeological excavations near Xi'an, where they viewed the "terra-cotta army"—7,000 life-size effigies of warriors, horses and chariots with individualized physical details—buried near the tomb of China's earliest emperor, Qin Shi Huang-di (210 B.C.). "It's surprising to realize that, even with all of the turmoil caused by the Cultural Revolution in China, their archaeological work didn't stop," notes Evelyn. "They have made amazing discoveries. We were very lucky to be among the first to see these sites."

Evelyn took one major trip in 1980—to the Himalayas, where she and Edward visited Nepal, Bhutan, Sikkim and Kashmir. The following several years afforded no time for travel because she and Edward dedicated themselves to the successful liquidation of the family business. After decades of frustration, they found that the "co-op mania" of the 1980s brought eager buyers for their properties. They seized this most advantageous moment to sell; the Borchard Real Estate interests had come full circle in three generations.

## Lucky in Life

It is undeniable that Evelyn Metzger has been lucky in life—she was blessed with a loving family, close friends, a luxurious upbringing, a happy and rewarding marriage and successful children. Perhaps luckiest of all, though, is that she has always had art as a lodestar to guide her. The pursuit of beauty—both that created by others and that which she herself has created—has provided intellectual stimulation, spiritual delight and a strong sense of purpose and motivation throughout her life.

Art, and the appreciation of it, constitute one of her principal legacies. Evelyn has dispersed the collection assembled by her parents, which she and Stuart later enhanced, to her children and to museums across the country. She has shared her enthusiasm for art at every opportunity. Edward recalls that museum visits have formed part of all of his trips with his mother since he was a child. "I remember especially visiting the Mauritshuis in The Hague, which houses Vermeer's *View of Delft*," he says. "Mother lingered interminably in that room. Finally she said, 'Eddie, look at this, contemplate it, study it—look at how perfect each brick is!'"

In a moment of philosophical reflection Evelyn sums up her views on the role of art in society. "Art is perhaps the noblest manifestation of a people. It is a distillation of their culture, the best and most enduring expression of who they are. Certainly, in my life," she concludes, "art has been paramount."

# Notes

p. 53
On one occasion, while making his way through the jungle, Ham suddenly found himself face to face with a puma peering down on him from a tree branch. Ham's only weapon was a pencil. Almost at once Ham and the puma got into a "stare you down" contest. Fortunately for Ham, the puma soon tired of this game, averted its eyes and wandered off.

p. 56:
Renowned art historian and critic Frank Jewett Mather, Jr. (1868–1953) studied at Williams College, Johns Hopkins University, the University of Berlin and the École des Hautes Études in Paris. He was director of the university museum at Princeton from 1922 to 1946 and the author of numerous books, including *Concerning Beauty* (1935), *Venetian Painters* (1936) and *Western European Painting of the Renaissance* (1939). Although trained to appreciate traditional realism, he was able to look upon the transition into modernism with an open mind. As a critic, Mather was more interested in placing an artist within a cultural context than in analyzing individual works.

p. 57:
In later years Mary and Evelyn resumed their friendship, lunching together and meeting at the Metropolitan Museum.

p. 68:
On a later trip to Egypt in 1949, Jefferson Caffery, their constant friend who was then ambassador in Cairo, insisted that they stay with him and his wife, Gertrude, at the embassy residence. Evelyn distinctly recalls an instance on this visit to Egypt when she was embarrassed by having to discuss the "art" of belly dancing. She was invited to tea by the aunt of king Farouk's wife, whose home was filled with her brother's paintings of belly dancers. Asked by her hostess what she thought of "our belly dancing," Evelyn sat in distressed silence momentarily before replying, "It must be marvelous exercise!" Only later did she learn that this was Egypt's national dance. In a similar vein, Gertrude Caffery was seated next to the Papal Nuncio to Egypt at a formal event when, as she said, "the nakedest tummy dancer" appeared. Gertrude clearly was shocked at this juxtaposition of nudity and Vatican propriety.

p. 68:
During a highly successful career, H. Freeman Matthews served as chief of mission in Stockholm, The Hague and Vienna. He was chargé d'affaires to the Pétain government and played an essential role in preparations for the allied landings in North Africa in 1942. Three years later he was on the diplomatic staff at the Yalta Conference during these crucial meetings between Franklin D. Roosevelt, Winston Churchill and Joseph Stalin. He was the first foreign service officer to attain the rank of Deputy Undersecretary of State (1950–53), becoming one of the first career ambassadors in 1956. Later, as ambassador to Vienna, he participated in the historic 1961 summit meeting between John F. Kennedy and Nikita Khrushchev.

In his memoirs, privately printed in 1973, Matthews vividly recalls his lengthy friendship with Ham and Evelyn Metzger:

> H. A. Metzger, who had been with Tropical at Barranca Bermeja for some years, came to take Palmer's place. He was extremely able, spoke Spanish like a native and proved to be an ideal representative of the company in those difficult years. He rose rapidly in the Jersey company and became their Latin American trouble shooter, serving from time to

time in Buenos Aires, Rio de Janeiro, Chile and their all-important Creole headquarters in Caracas. Also, he was a great help to the Chief (Jefferson Caffery) and me; he knew and kept us informed of all oil developments and government moves and plans. In addition he was—and is—delightful company, with a keen sense of humor, and is an amusing raconteur. Many of his tales were at his own expense for he was modest and devoid of any sense of pomposity. He is just my age (Cornell '21) and has remained one of my closest friends—and also of the Chief—throughout the years; for a brief period it seemed likely he would also be my brother-in-law. However, in the '30s he married Evelyn Borchard, a delightful, charming person and an artist of international repute. We always see them when we visit New York. (p. 88)

p. 85:
Between 1966 and 1973, when Spruille was president of the Metropolitan Club, Ham and Evelyn were often his guests at events, including the yearly Easter Sunday buffet.

p. 88:
William Leonhart received an M.A. from Princeton in 1941 and a Ph.D. in 1943. Selected as special assistant to Nelson Rockefeller, then Coordinator of Inter-American Affairs at the Department of State, he entered the foreign service in 1943 as an economic analyst stationed in Buenos Aires. His career included postings in Belgrade (1946), Rome (1949), Saigon (1950) and Tokyo (1951, 1958). He became ambassador to Tanganyika in 1962. He was detailed to the White House in 1966, also becoming career minister and receiving the Superior Honor Award in the same year. In 1967 he was made special assistant to the president on Viet Nam, civilian programs. In 1969 he was assigned as ambassador to Belgrade. Two years later he was detailed to the commandant of the National War College.

p. 96:
In fact, Helena's husband arranged to have Franco intercede with Perón, who ordered Francisco Cambó's collection returned to Spain. Interestingly, while Helena was an only child, she and her husband were to have fourteen children.

p. 97:
The wealth of Argentina derived in large part from the raising of cattle and horses on vast estates worked by gauchos. Over time many of these estates came to include luxurious residences or estancias, with large recreational facilities where the world's best polo players developed their skills. When the Mihanovich family had the Metzgers to their estancia, they presented two of their sons who were then world-champion polo players.

p. 98:
Joseph Page describes the burning of the Jockey Club as follows:
> Three jeeps came to the Tucumán Street entrance. The majordomo tried to keep them out, but they forced their way by him. Several men with hatchets went upstairs to the main hall and hacked original paintings by Goya and Velásquez out of their frames. They piled them on the floor and started a bonfire. They also smashed to pieces a statue of Diana the Huntress. The majordomo called the fire department. The answer he received was "We have no instructions to put out a fire at the Jockey Club." The burning of the Jockey Club amounted to one of the worst instances of art vandalism ever perpetrated. (p. 272)

p. 98:
Once back in the United States, Jim went to Choate, Ed to Lawrenceville (later to Collegiate) and Eva to Brearley. Jim went on to receive a B.A. from Cornell and an M.B.A. from Columbia's Graduate School of Business. Ed graduated from Hobart College and Eva also graduated from Cornell.

p. 98:

Mercedes Cambó, who had been a devoted friend and had extended her extraordinary hospitality to all the Metzgers throughout their stay in Argentina, accompanied them on the first leg of their trip, to São Paulo, Brazil.

p. 102:

In May 1990 the Leonharts invited Evelyn to the opening of the exhibition *Yokohama: Prints from Nineteenth-Century Japan* at the Arthur M. Sackler Gallery of the Smithsonian Institution in Washington. The gallery's director, Milo Cleveland Beach, wrote in the foreword to the exhibition catalogue:

> The period immediately following Commodore Perry's voyage is the subject of this exhibition of prints from the collection of William and Florence Leonhart of Washington, D.C. The works, which Ambassador and Mrs. Leonhart assembled during Mr. Leonhart's tours of duty in Japan, illustrate the Japanese response to the new peoples and ideas to which they were introduced in that eventful midcentury era. They also embody the Leonharts' serious interest in Japan and the United States.

The prints are universally fascinating, and some are very amusing. Images from the Leonharts' collection also are reproduced on the cover and in the lead article ("After Centuries of Japanese isolation, a fateful meeting of East and West: Japan's Opening to the West") of *Smithsonian* 25, no. 4 (July 1994): 20–33.

p. 103:

In addition to George and Alice, the Browns' party included their daughters Maconda O'Connor and Nancy Negley (there with her husband), Helen Daniels Allen (whose husband Herbert Allen was president and chairman of Cameron Iron Works, Inc.) and Oveta Culp Hobby, chairman and editor of the *Houston Post.*

p. 110:

Ham Metzger's business papers are now at the Center for Business History Studies at Tulane University. Exxon Corporation issued an official press release on Ham's death October 21, 1974, which stated, before tracing his career, "Mr. Metzger was one of the best known figures in business and financial circles of the hemisphere. He was active in many Latin American-oriented organizations." *Who's Who in America* (1961) and *World Who's Who in Commerce and Industry* (1961) state that Ham was a member of the National Industrial Conference Board, the Council on Foreign Relations, the Pan American Society, the Foreign Policy Association, the Venezuelan Chamber of Commerce Club and the Economic Club (New York City).

In providing biographical data to Tulane University, Evelyn followed her summation of Ham's career with some personal observations about her husband. "He was a warm, happy man, who enjoyed every phase of life: his work, recreation, his family, his friends . . . . He was scrupulously truthful . . . accurate and dependable. He was witty, but not abrasive; uncomplicated, but aware of subtlety . . . . He was gregarious, but could be content alone. In spite of his self-sufficiency, he was a team player. . . . Ham often said that if he had his life to live over again, he would gladly spend it the same way."

p. 111:

Alice Fordyce, whom Evelyn knew as a member of the Waltz Group, went to Evelyn's apartment to brief her on the 1979 trip to China. Alice and her sister, Mary Lasker, had been in China even earlier as guests of the Chinese government.

# Bibliography

Amory, Cleveland. *The Last Resorts.* New York: Harper, 1952.

Bauld, Harold J., and Jerome B. Kisslinger. *Horace Mann-Barnard: The First Hundred Years.* New York: Horace Mann-Barnard School, ca. 1986.

*The Bohemian Watering-Places and Their Environs: Guide and Hotel Registry.* Karlsbad: Für Fremdenverkehr in Dëutschböhmen, 1909.

Brockway, Wallace. *The Albert D. Lasker Collection: Renoir to Matisse.* Private printing.

Carson, Gerald. "New York in the 20s." *Amercian Heritage* (November 1988): 45–82.

Crassweller, Robert D. *Perón and the Enigmas of Argentina.* New York: Norton, 1986.

Eells, George. *The Life That Late He Led: A Biography of Cole Porter.* New York: G. P. Putnam's Sons, 1967.

Fitzgerald, F. Scott. "Echoes of the Jazz Age." In *The Crack-Up,* ed. Edmund Wilson. 1945. Reprint. New York: New Directions, 1956.

Gallagher, Brian. *Anything Goes: The Jazz Age Adventures of Neysa McMein and Her Extravagant Circle of Friends.* New York: Time Books, 1987.

Gibb, George Sweet, and Evelyn H. Knowlton, *History of Standard Oil Company (New Jersey): The Resurgent Years, 1911–1927.* New York: Harper, 1956.

Goldsmith, Barbara. *Little Gloria . . . Happy at Last.* New York: Alfred Knopf, 1980.

Greenup, Ruth, and Leonard Greenup. *Revolution before Breakfast: Argentina 1941–1946.* Chapel Hill: University of North Carolina Press, 1947.

Gunston, David. "Master of the Pearl: The Story of Kokichi Mikimoto." *Oceans* 37, no. 3 (May–June 1984): 17–21.

Harris, Corra. "How New York Amuses Itself." *The Independent,* 16 March 1914.

Hochfield, Sylvia. "Lydia Winston Malbin: A Futurist Eye," *ART News* (April 1988): 91–92.

Joseph, Jonathan J. *Jane Peterson: An American Artist.* Boston: private printing, 1981.

Loti, Pierre. "Impressions of New York." *Century Magazine* (February/March 1913).

Maher, James. T. *The Twilight of Splendor: Chronicles of the Age of American Palaces.* Boston: Little, Brown, 1975.

*Masterpieces of Art: In Memory of William R. Valentiner, 1880–1958, Representing His Achievements during Fifty Years of Service in American Museums.* Raleigh: North Carolina Museum of Art, 1959.

McIver, Stuart B. *Yesterday's Palm Beach.* Miami: E. A. Seemann Publishing, 1976.

Morris, Lloyd. *Incredible New York: High Life and Low Life of the Last Hundred Years.* New York: Random House, 1951.

Nadelhoffer, Hans. *Cartier: Jewelers Extraordinary.* New York: Abrams, 1984.

Page, Joseph A. *Perón: A Biography.* New York: Random House, 1983.

Pratt, Theodore. *That Was Palm Beach.* St. Petersburg, Fla.: Great Outdoors Printing, 1968.

"Royal Visitor Tops Off a Strenuous Day with Attendance at Opera." *New York Times,* 19 Nov. 1919, p. 3.

Runyon, Damon. *Runyon First and Last.* Foreword by Clark Kinnaird. Philadelphia and New York: J. B. Lippincott, ca. 1949.

Sterne, Margaret. *The Passionate Eye: The Life of William R. Valentiner.* Detroit: Wayne State Univ. Press, 1980.

*Time Capsule/1923: A History of the Year Condensed from the Pages of Time.* New York: Time Inc., 1967.

Trager, James. *West of Fifth: The Rise and Fall and Rise of Manhattan's West Side.* New York: Atheneum, 1987.

Wall, Bennett H., and George S. Gibb, *Teagle of Jersey Standard.* New Orleans: Tulane University Press, 1974.

# The Rich Artistry of Evelyn Metzger

NANCY G. HELLER

"YOU DO WHAT you have to do, and I have to paint."[1] Indeed, Evelyn Borchard Metzger has devoted her life to making art. For more than seventy-five years, despite the demands of her roles as wife of a corporate executive and mother of three active children; despite the distractions involved in making numerous international moves; and despite pressures from people who felt it improper for a woman to regard her own work as anything more than a hobby, Metzger has always remained a serious painter.

The result of this single-minded devotion is an artistic oeuvre as remarkable for its diversity as for its size. Through the years Metzger has explored many different subjects and styles, including academic realism, Impressionism, Neo-Impressionism, Fauvism, Cubism and pure abstraction. Likewise, she has tried her hand at a variety of artistic techniques, from charcoal and pen and ink drawing to watercolor, oil on canvas or board, poured enamel, spray-painting and the use of stencils and collage. She also has produced monotypes and portrait sculpture. Metzger's accomplishments are impressive. Since her first one-person exhibition in Argentina in 1950, she has had seventeen additional solo shows—in Mexico, France and all over the eastern and midwestern United States. She has participated in numerous group exhibitions; has work in the Department of State's Art in Embassies program; and is represented in the permanent collections of important art museums in more than twenty states.

At first glance, much of Metzger's art, while attractive and skillfully composed, seems to have little to do with the realities of human existence. A closer look, however, reveals that Metzger's best works contain an emotional subtext, a subtle sense of melancholy, sometimes even alienation, that makes these pictures especially affecting. Like the early 20th-century American master Edward Hopper, whose work she admires, Evelyn Metzger creates deceptively simple artworks that are often "about" far more than their ostensible subjects.

## A Cultural Upbringing

Evelyn Borchard Metzger and her older brother, Stuart, grew up in New York City during the 1910s and 1920s in a spacious, six-story structure on Manhattan's Upper West Side. Their father, Samuel Borchard, was a successful factory owner who soon made his mark in New York real estate. A highly intelligent man, Borchard expected his children to excel academically—which they did. Evelyn graduated Phi Beta Kappa from Vassar College in 1932. He was also directly responsible for her interest in golf (her trophies fill several shelves in her library). Eva Rose Borchard, Evelyn's mother, was considerably younger than her husband. A strikingly handsome woman, she is described by her daughter as having intuitive good taste and a tremendous zest for living.

By example, Evelyn's parents taught her that art—specifically the experience of studying great original artworks in person—was an essential part of life. They took the children with them on their annual summer trips to Europe, where tours through the major art museums formed part of each day's itinerary. Her parents were also art collectors, assembling an impressive array of Old Master European oil paintings—primarily by Italian Renaissance and Dutch 17th-century artists. Since a number of her parents' friends were also collectors, Evelyn grew up in a world where it was normal to have a Bellini in the parlor and a Brueghel in the hall. In addition, Samuel and Eva were careful to include both children, beginning at the age of ten, in their Sunday luncheons—salons, really, to which distinguished figures in many fields were invited. It was during these Sunday events that young Evelyn became acquainted with many of the most important figures in art scholarship. She met such people as Karl Lilienfeld, an expert on 17th-century Northern European paintings, who taught her to study pictures at close range, to be able to see the artists' brushstrokes or "handwriting." Another luncheon guest was William Valentiner, a noted art historian, editor and museum director, who wrote monographs on Rembrandt van Rijn, Frans Hals and Pieter de Hooch. During a family visit to Berlin, Evelyn, aged ten, met another significant scholar, Wilhelm von Bode. A museum director and renowned expert, Bode was the author of seminal volumes on Italian, Flemish and Dutch painting.[2] This personal contact with some of the most illustrious connoisseurs of the day clearly whetted Evelyn's appetite for further study, inspiring her own art-historical writings.[3] It also gave her an unusually strong background for creating art.

## Early Training

Evelyn was passionate about visual art well before she began attending the family's Sunday luncheons. She still vividly recalls the frustration and envy she felt, as a four-year-old, when her brother was given his first set of drawing pencils, and she was left out. Three years later her parents redressed this inequity, enrolling both children in classes at the Art Students League.

Founded in 1875, the league has always been unique—an inexpensive, coeducational, student-run institution that offers daily drawing classes, along with critiques by noted artists twice a week. An extraordinary cross section of painters has been associated with the Art Students League, ranging from Thomas Hart Benton to Thomas Eakins, Helen Frankenthaler to Audrey Flack and Georgia O'Keeffe to Norman Rockwell.

Evelyn and Stuart were the youngest members of George Bridgman's Antique Class at the league. There, following the centuries-old tradition still popular in academic art schools today, students learned to draw by painstakingly copying plaster casts of well-known sculptures from ancient Greece and Rome (thus, the term *antique*). Since virtually all the statues copied represented the idealized human figure, drawing from casts was intended to give artists an appreciation of ideal beauty, a thorough command of human anatomy and the skill to render accurately three-dimensional forms on a two-dimensional surface. Bridgman, a man in his fifties, was just beginning what would become a highly successful thirty-year teaching career at the league.[4]

Unfortunately, no examples remain of Evelyn's earliest drawings from the Art Students League. As she tells the story, her mother thought they were the teacher's work and threw them out. She still has pictures dated just a few years later, however—an album filled with small-format, black-and-white drawings she made between roughly the ages of ten and twelve.

# First Works

A surprisingly large amount of information can be deduced about the artist as a young girl, from these informal, often amusing, images. One of the earliest pictures in the group is a drawing of couples enjoying themselves at a tea-dance held at the Coconut Grove of the Royal Poinciana Hotel (fig. 12). This was an elegant establishment in Palm Beach, Florida, where the Borchard family wintered during the later 1910s and 1920s. Evelyn still recalls watching the dancers from her hotel window. While clearly the work of an unformed artist, even this picture reveals her sophisticated sense of composition—balancing the wooden structure holding lamps on the far right against a large expanse of empty space on the opposite side—and her Impressionistic interest in depicting pleasant subjects, caught at a single moment in time. Moreover, the technique she has used here, and in all these early pictures, is both unusual and intriguing. By covering almost the entire surface of each picture with rich, black ink, so that the narrow white lines appear to have been scratched through the dark areas, the artist has created a heightened sense of drama.

96
*Fantasy Man*
ca. 1921
Ink on paper
5⅝ × 3½ in.

*97*
*Inkwell Nude*
ca. 1921
Ink on paper
9½ × 7½ in.

Evelyn's early drawings reveal an active imagination, especially, in a series of grotesque heads, stippled with complex, delicate patterns of short lines. Some of them are further embellished by elaborate, Art Deco-style corner designs.[5] Even the more conventional drawings of female subjects from this period are wildly stylized and rather emotional. The "flapper" (fig. 14), seen in profile, has her lips parted and her eyes open wide, as though focusing on some shocking scene just beyond the upper right-hand corner of the paper.[6] Another woman, whose face, body and accessories have all been reduced to a series of triangles, looks tired, or perhaps unhappy (fig. 15). There is also a drawing of an almost robotic-looking male with a curiously up-to-date "Mohawk" haircut (fig. 96). One of the most interesting pictures from this group, however, features the gentle, romantic image of a slim young woman with her hair pulled back into a low bun, wearing an elegant evening dress with a shawl draped seductively across one shoulder (fig. 16). Judging from the photos in her high-school yearbook, this would seem to be a generalized self-portrait of the artist. On the other hand, Evelyn's drawing of a nude woman, seen from the rear, holding a quill pen in one hand and a scroll in the other (fig. 97), is reminiscent of a study from the antique or model from a life-drawing class. In fact, it is based on an Art Deco brass inkwell owned by her family, although, ironically, it also bears more than a passing resemblance to the official seal of the Art Students League.

Another example of Evelyn's extant juvenalia is in color. In 1928, after spending a number of seasons in an apartment building near the Royal Poinciana, the Borchards purchased a house in Palm Beach built by the celebrated architect Addison Mizner. Evelyn painted a lively portrait of it, using thick paint and energetic brushstrokes, on the outside of a tin box (plate 1).

## The Horace Mann Experience

Once Evelyn had completed her studies at Robert Louis Stevenson, which she describes simply as "a local neighborhood elementary school," her parents enrolled her at the Horace Mann School, long noted for its high academic standards. Founded in 1887 as an experimental unit of Columbia University's Teachers College, Horace Mann High School for Girls was located at 120th Street and Broadway, about thirty blocks north of Evelyn's home.[7] While art was not part of the school's curriculum during Evelyn's tenure there, she did belong to an art club called the Paint Pot. She also served as an art editor of the school's yearbook.

Of the forty-seven people in Evelyn's graduating class, at least one other woman also became a professional artist. Recently Eline Holst McKnight—a printmaker, best known for her woodcuts—was kind enough to share her high school memo-

---

*"Evelyn was one of the smartest girls in our class, and very beautiful. She was fluent in French, too—I used to envy her."*

98
The Paint Pot Club at Horace Mann School for Girls with Evelyn Borchard seated third from right, second row. Collection Alumni Office, Horace Mann-Bernard School.

ries of Evelyn. "I've always admired her [artistic] gift," McKnight said; "Evelyn was one of the smartest girls in our class, and very beautiful. She was fluent in French, too—I used to envy her."[8] The two women lost touch with each other over the years. In the mid-1980s McKnight saw her classmate again, and for the first time got a look at her mature artworks. She said she was amazed at the sheer volume of Metzger's output, noting particularly "her excellent draftsmanship—she was obviously thoroughly trained—and her beautiful colors."[9]

## The Vassar Years

In 1928 Evelyn Metzger first began her formal study of art history, at Vassar College. There she was greatly influenced by Agnes Rindge, with whom she took several art history courses. "She really made me work," Evelyn recalls. "She gave me my only C in college—in Baby Art [the introductory survey course]—to stop me from being too cocky; I thought I knew it all." Some of Evelyn's youthful arrogance is understandable, since she was reading about artworks she had already studied in situ. Still, she cites Rindge as a formative influence, and calls her "a great teacher . . . who had original reactions to art, saw things differently" and encouraged her students to think seriously about what they observed.[10]

Another important part of her Vassar education was provided by Clarence (C.K.) Chatterton, who taught painting. Like other traditional educational institutions of the 1920s, Vassar emphasized academic rather than practical courses, so Evelyn's time in the studio was restricted (fig. 99). Still, she appreciated Chatterton's teaching style, which she describes as "very encouraging." At the end of her last year at Vassar, Chatterton's encouragement, combined with her own spunk, raised more than a few eyebrows. Because she was so strong in figure-painting, Chatterton suggested that Evelyn paint a full-length, life-size, female nude. She did, and the result was so successful that it was hung prominently in an end-of-term exhibition. A number of the Vassar alumnae who visited the show were shocked, however. They complained to Oliver Tonks, chairman of the Art Department, who responded by removing the painting from public view and placing it in his own office.

Evelyn's father died during her sophomore year at college. Uncomfortable staying in the house near Riverside Drive, Mrs. Borchard moved the family into the newly completed Waldorf Astoria Towers in 1931. As Evelyn recalls it, the Borchards' apartment on the thirty-third floor was "beautiful, with a magnificent

99
Vassar Art Department - Class in Life Drawing, late 1930s, by Margaret M. DeBrown. Courtesy Special Collections, Vassar College Libraries.

Plate 10
*Model Knitting
with Artist at Work*
n.d.
Oil on panel
32 × 24½ in.

Plate 11
*Model in Yellow Turban*
n.d.
Oil on panel
24 × 17¾ in.

100
Eva Metzger with her portrait bust, ca. 1952.

---

*"By the age of twelve I had a knack for likenesses, and when you can do something, you tend to like it."*

view." This elevated vantage point may have had some influence on the aerial views of Central Park and other landscape subjects which the artist later painted. She would also have known that such angles were popular with several French Impressionist painters—notably Claude Monet and Camille Pissarro—who often depicted Parisian street scenes observed from a hotel window.

## Figure-Painting and Sculpture

Freed from the demands of college classwork, Evelyn went back to painting seriously after her Vassar years. Between 1932 and 1934, she periodically returned to the Art Students League, for a month or so at a time, where she worked with the German Expressionist George Grosz and the American Social Realist Raphael Soyer. While none of Evelyn's art is either savagely distorted, like Grosz's work, or straightforwardly critical of American society, like Soyer's, something of their sensibility can be seen in Evelyn's early South American portraits, in which she conveys a sense of empathy with the impoverished local residents whose images she has painted.

She also spent time, on and off, at the National Academy of Design. The academy was older than the Art Students League, having been founded in 1825, and more conservative. It offered numerous life classes (in which nude models could be painted or drawn) at a more convenient location, however. As Evelyn says, she went to the academy "for the models." Few of her works from this period can be securely dated, but we get a sense of the kind of paintings she was producing through a set of academic nude studies of seated women—one knitting, one posing while wearing a big picture hat and the third, in a magnificent yellow turban, resting her chin on her hand. The sense of space created in the picture with the knitting model (through the depiction of another painter set further back) is quite believable (plate 10). The solidity and weight in the turbaned figure's eyelids and elbow demonstrates that Evelyn had developed a secure command of pictorial structure (plate 11). Moreover, even though these are clearly classroom studies, with the models in standard poses, the paintings also illustrate the cool, emotionally distanced attitude Evelyn takes toward most of her subjects. The gazes of two of the women are directed down or sideward, avoiding the viewer's glance. The face of the third model is treated in such a loose and generalized manner that it is impossible to tell where she is looking.

It was also during this time that Evelyn did some of her most intensive work in sculpture. Along with her periodic visits to the Art Students League and the

*Plate 12*
*Portrait of Jefferson Caffery*
1940
Oil on canvas
26 × 21 in.
Collection The Jefferson Caffery Papers, Southwestern Archives and Manuscripts Collection, University of Southwestern Louisiana

National Academy of Design, she spent some time during the early 1930s studying with Sally Farnham, a prominent New York sculptor best known for the bronze equestrian portrait of the Venezuelan liberator Simón Bolívar (1921), located at Central Park South and 59th Street.[11] Evelyn recalls that she went to Farnham's studio on Central Park West a few times a week. While she quickly realized that she would only have time to pursue a career in either painting or sculpture, ultimately opting for the former, nevertheless this experience was significant. The specific influence of Farnham is evident in the busts Evelyn made of her children and husband while living in Argentina. Like Farnham's work, Evelyn's portraits demonstrate an ability to catch striking likenesses of her subjects, and a meticulous attention to detail.[12] The bust of daughter Eva as a child, in particular (fig. 100), shows her mother's versatility—working successfully in both three dimensions and two.

In 1933 Jane Peterson, a well-known painter and a family friend[13], decided that Evelyn should meet Herman A. Metzger (widely known as "Ham"), who was visiting New York. A tall, handsome man in his early thirties who had taken his engineering degree from Cornell directly to Latin America, Metzger was already well established as an executive with Standard Oil Company (New Jersey) in Bogotá, Colombia. As soon as they met, at one of Eva Borchard's cocktail parties, the young people were smitten. Evelyn recalls, "We went out three nights in a row—and then he had to go back to South America." The relationship continued, via long-distance mail, and the following year they were married. The couple honeymooned in Europe and then settled in Bogotá. Since the two of them were fascinated by new places and new experiences, Evelyn notes, "We had the best of both worlds—he had never been to Europe before, and I had never been to South America." Thus began a new chapter in her life, as a married woman and an American living abroad.

## *Early Sojourn in South America, 1934–44*

Although she describes herself as rigid, in fact Evelyn Metzger has been remarkably flexible, adjusting to all sorts of changes throughout her life. During the first years with Ham this skill was sorely tested: in one three-year period, they lived in eleven different homes. Altogether, during the course of two decades in South America, the Metzgers resided in five countries, living at twenty different addresses in as many years. This gave them the dubious distinction of changing

*101*
*Bogotá Resident*
ca. 1935
Oil on canvas
20 × 18 in.

*102*
*Bogotá Resident*
ca. 1935
Oil on canvas
20 × 18 in.

United States Legation in Colombia[15] and a neighbor of Ham's. It was natural for these two American bachelors to strike up a friendship. In the years before Evelyn's arrival on the scene, they often went hiking and mountain-climbing together on weekends. When both men married, they continued their friendship as a foursome. Metzger's picture (painted in 1940, when the Metzgers were residing in Brazil, where Caffery was then ambassador) emphasizes his head, with much more attention given to the planes of his face and the details of his features than the relatively summary upper body (plate 12). Working in the tradition of 17th-century European portraiture, Metzger has made her sitter appear relaxed and lively, through the off-center placement and slight angle of his head, plus the hints of a furrowed brow and pursed lips. Like subjects portrayed by Hals or the Italian Baroque sculptor Gianlorenzo Bernini, rather than simply posing, Caffery appears to be thinking about something and preparing to respond to the artist.

In addition to portraits, Metzger also painted some landscapes during these years. One of the most interesting examples depicts the northern Colombian seaport of Cartagena (fig. 104). Painted from the top of the six-story Andian Building, then the tallest modern structure in the city, this work features an unexpected viewpoint—raking across roofs and catching only the tufted tops of the palm trees. The oil tanker in the distance was added by Ham in his only direct contribution to his wife's oeuvre. She also painted a Cubist variation on the same scene (fig. 105). It is interesting to compare the two pictures, which provide almost a textbook demonstration of the process of simplifying, and stylizing, a given subject. The Cubist Cartagena has lost remarkably few of the details included in the more realistic version—one can still identify the palm tree in the center of the composition, for example, and the odd, saw-toothed top of the building in the lower left. As the forms have been flattened out by the removal of highlights and shadows, the sense of real space has been destroyed, however, and a more decorative surface pattern created.

In 1938 the Metzgers relocated to La Paz, Bolivia, where they lived periodically for three years. Between this time and 1944, when the family moved to Argentina, Metzger's artistic output diminished considerably. This was largely due to two factors: their peripatetic lifestyle (living in Rio de Janeiro from 1940 to 1941, Santiago de Chile for the following thirteen months and then—briefly—back in Rio) and the births of their children.

At the same time, Metzger clearly relished the excitement of exploring new places. She remembers, for example, the brilliant colors and strong sunlight of La Paz, where she painted the marketplace and local Indians in traditional dress (plate 13). Also in La Paz, she got together with a group of artist-friends and invited a noted Bolivian painter, Guzmán de Rojas, to give periodic critiques of their work. In Peru, where she and Ham did a great deal of sightseeing, Metzger was particularly impressed by the ruins from the Inca civilization. Ruined stone structures—whether in South America, Egypt or Cambodia, have always held a special fascination for Metzger; they figure prominently in both her paintings and her husband's photos, for which Evelyn acted as "art director."

The stunning black-and-white photographs that Ham Metzger took during these years provide a tangible record of the family's travels to many of the Western world's most breathtaking sites, at a time when modern urbanization, industrialization and commercialization had not yet taken hold. They also reveal some striking similarities to Evelyn Metzger's paintings of related subjects. For example, like his wife's painted portraits, Ham's photos of South American Indians lovingly present each pattern in their intricately woven shawls and every tassel or other bit of decoration on their remarkable hats, without sentimentalizing his subjects or relegating them to the status of the picturesque. These are powerful people, whose unexpected silhouettes have been juxtaposed, matter-of-factly, against examples of their local architecture or extravagantly beautiful natural landscapes. The portrait photos, like Metzger's painted counterparts, set their human subjects close to the viewer, while landscape views stress diagonal lines—created by cobbled roads, railroad tracks or arcades—that draw the viewer deep into each image. Another similarity is the effective use of dramatic shadows, by both photographer and painter.

## The Buenos Aires Decade, 1944–54

After so many years of near-constant packing and unpacking, it must have been a relief to stay in the same city for an entire decade. Of course, it also helped that the city in question was such an intriguing one: Buenos Aires, where the Metzgers lived from 1944 to 1954.

This was a critical period in Evelyn Metzger's life, both as a woman and an artist. The Metzgers remained in Buenos Aires long enough to become well acquainted with many of the city's writers, artists, businessmen and politicians. Perhaps because the family remained relatively settled, in Buenos Aires Metzger produced a large number of paintings—mostly landscapes and portraits, a remarkable two hundred or more of the latter. During these years she experimented with a new color system and studied, via the same informal method she had experienced in Bolivia, with the well-known Argentine painter Demetrio Urruchúa. It was also in this city that Metzger realized one of the professional artist's most important goals: her first solo exhibition.

Although they live in the capital of the second-largest country in South America, *porteños,* as the residents of Buenos Aires are called, have always considered themselves European. Indeed, their city is nicknamed "the Paris of South America," because of its wide boulevards, large public parks and ornate buildings, many of which have a distinctly French appearance. A French influence can also be detected in some of Metzger's landscapes.

The urban and rural landscapes Metzger painted during her years in Argentina clearly demonstrate the ways in which much of her work resembles standard 19th-century French Impressionism (with its informal looking, momentary glimpses of pleasant places and attractive, upper-class people) and also the ways in which it departs from that model. Diverging from the Impressionists, Metzger uses black extensively and creates forms that appear solid, rather than ephemeral. She also employs compositional and other devices that distance the subjects from the viewer, both physically and emotionally. Metzger's Argentine views are extremely attractive—loosely brushed canvases featuring a palette of dusty blues, yellow-greens and rose—very much like the colors found in a box of pastel chalks and on the exterior walls of much Central and South American architecture. These pictures have an easy-going air about them and a marvelous command of the curved line. One arresting example of this type shows three horse-drawn carriages lined up at an urban corner, the horses resting as their drivers wait for customers (plate 5).

The subject of carriages is obviously an important one for Metzger, as it recurs periodically throughout her oeuvre. ("I was always fascinated by carriages," she says, "in Italy, Buenos Aires, wherever.") Here, the carriages lend a characteristic sense of romance to the city scene. As one of Metzger's friends from this period said, "Buenos Aires is so old and historic—the city looks nostalgic!"[16] More specifically, these carriages remind Metzger of a particularly enjoyable evening in New York, when she and Ham were courting.[17]

The odd perspective going around the corner in this work, and the cockeyed angle of the pole on the far right of the canvas, are further evidence of Metzger's intentional distortions of reality. The vast expanse of empty roadway dominating this picture's foreground is presumably a nod to the sections of deserted Parisian streets painted by the French Impressionist Gustave Caillebotte and the empty

ballet-studio floors of Edgar Degas. The solidity of the architectural planes, however, along with the flattening out of the horses' bodies—all painted essentially one shade of a given color—and the heavy, crudely painted black outlines of the forms, give this picture a style all its own.

Metzger's exterior view of the church of San Isidro (plate 6)—located in the popular coastal town, some ten miles from Buenos Aires, where the wealthy had their summer villas—was presumably one of her favorites, as it appears on the cover of the announcement for her first solo show. Here, a lone man rests on a park bench outside the church grounds, facing away from the viewers. So positioned, he simultaneously arouses our curiosity about what he is thinking and draws our gaze deeper into the picture. Metzger has used such thick pigment in some places, and applied it with such vigor—often using a palette knife—that it is actually difficult to focus on the objects portrayed, rather than the abstract forms created by the swirls of paint. In this work, as with the carriages, Metzger reveals her skill at painting trees—which would remain one of her strongest subjects for decades to come. Particularly intriguing is the way the bare branches of the tree at the far right interact with the sections of blue-gray sky they frame. As in some of the Dutch painter Piet Mondrian's early treescapes, here, too, it is often

*Plate 14*
*Windsor Chair, Lytton Lodge on Lake Mooselookmeguntic*
ca. 1949
Oil on canvas, 23 × 19 in.

*106*
Vincent van Gogh
*Gauguin's Armchair, Candle and Books*
1888
Vincent van Gogh (Stichting/Foundation)/Van Gogh Museum, Amsterdam

difficult to tell which forms are supposed to be read as "positive" (solid), and which "negative."

The most effective of all Metzger's Buenos Aires paintings is the simplest, based on another subject from outside the city: the trees at Llao Llao. Although the Metzger family spent little time traveling outside Latin America during the late 1940s and early 1950s, they did go to considerable lengths to explore Argentina itself at a time when travel within the interior was far less common, comfortable or safe than it is today. In her 1944 "Letter from Buenos Aires," published in the *New Yorker,* the Metzgers' friend Jean Boley writes that, traditionally, porteños had not been interested in visiting the rest of their country, preferring to vacation in European capitals and resorts.[18] Boley adds, however, that with the increasing practical difficulties caused by World War II, a "See Argentina First" movement began to have some effect. She specifically mentions the growing curiosity among local tourists about the lake district in southwestern Argentina. As Boley points out, such places were not yet generally popular, because they offered few paved roads for automobiles and the train trips were quite long and uncomfortable. Thanks to the Standard Oil company plane, the Metzgers and their friends did investigate these resort areas, though, visiting Bariloche on trout-filled Lake Nahuel Huapi and Llao Llao, a nearby village that featured its own port. Metzger still recalls the dramatic views they saw in Bariloche: "There are volcanic mountains and blue, blue lakes, very deep. . . . They have good fishing there, beautiful flowers and trees."

Her painting of Llao Llao (plate 4) contains a number of distinctly Fauve elements—most notably, the manner in which each rounded section of earth is painted a different, vivid and entirely arbitrary color.[19] Like the other Argentine landscapes discussed here, this painting has strong black outlines around every form and ambiguous spatial effects caused by the interplay of the curving branches and the planes beyond them. In the distance there is also the profile of a grayish mountain—reminiscent of Paul Cezanne's paintings of his family's estate in the south of France (fig. 78) and Mont Sainte-Victoire series.

During these years, on periodic trips to the United States, Metzger also made numerous drawings of the interior of her family's Maine property—seven buildings by Lake Mooselookmeguntic, all filled with early-American-style furniture and antiques. Another work of this vintage is an oil painting focusing on a single Windsor chair (plate 14). This almost iconic representation of the empty chair is oddly arresting. Placed dead-center within the composition, assertively facing the viewer with its perspective deliberately skewed, the chair has rich colors and lushly painted outlines.

Characteristically, Metzger denies that there is any significance inherent in this subject. When asked why she chose it, the artist says, simply, "It was there, in Maine, set against the view of the lake, and it was *paintable*." Metzger asserts that the chair holds no personal associations for her or the other members of her family. It was no one's favorite chair, she says: it was simply one of many pieces of furniture the family had to dispose of, when they donated the Maine camp to the Boy Scouts of America. This is more than a conventional interior view, however. Rather, it is a portrait of the Windsor chair, which is given the same kind of importance, and individuality, that one of Metzger's human subjects might have. Further proof of the resonance this chair holds for Metzger is the fact that she still has it—in the library of her Manhattan apartment. Significantly, she notes that this is the only object she has retained from the family's Maine property.

In some ways, Metzger's painting calls to mind the evocative representations Vincent van Gogh made of his own favorite chair and the one Paul Gauguin used when visiting the Dutch painter in southern France (fig. 106). Although the Dutch painter's canvases are more brightly colored and filled with aggressively swirling brushstrokes, those, too, are centralized and iconic, depicting inanimate objects that seem to be imbued with the spirits of their users. Metzger's *Windsor Chair* also looks ahead, to the strong personalities with which she would imbue so many of the trees she painted in her Central Park series, years later.

## Art and Flowers – First Solo Exhibition

Although many of her most effective works are landscapes, the bulk of Evelyn Metzger's oeuvre from the Argentine period is made up of portraits. Since childhood, she has enjoyed painting people's likenesses and has had success with them. As Metzger notes, "By the age of twelve I had a knack for likenesses, and when you can do something, you tend to like it." Just how good she is at representing her sitters, in both two and three dimensions, can be seen by comparing the portrait bust she has created to the subject seated next to her backyard worktable, in an earlier photograph from her Palm Beach years (fig. 34). As Metzger says, looking back on the body of her work, "By and large . . . I'm a realist. I spent considerable time on the abstractions [of the 1960s] . . . but I'd always go back to the realistic subjects, and portraiture."

*107*
Eleanor Mallory
ca. 1945
Oil on canvas
18 × 14 in.

*Plate 15*
Jean Boley
1946
Oil on canvas
26 × 21½ in.

*Plate 16*
Lucy Burrows
1947
Oil on canvas
19 × 16 in.
Collection Mrs. Charles R. Burrows

*Plate 17*
*Ed and Jim*
ca. 1946
Oil on canvas
21½ × 24¼ in.

*Plate 18*
*Jim Reading*
ca. 1947
Oil on panel
27 × 19½ in.

Her artistic output during these years seems extraordinary, until one realizes what a disciplined and dedicated worker Metzger always has been. Despite her considerable social obligations—as both hostess and guest—and the pressures of running a busy household, Metzger spent as much time as possible in the family's suburban Buenos Aires home, painting. Because it quickly became fashionable to have her execute one's portrait, Metzger eventually painted "all the embassy crowd," plus a great many other friends, visitors and family members.

Most of these portraits are fairly formal, bust-length and relatively still, with their subjects carefully coiffed and dressed. Several of these pictures were painted using what Metzger refers to as her "limited-palette" technique, an experimental system in which every part of the painting is made using only four colors—black, white, burnt sienna and yellow ocher. The challenge, of course, is to see how broad a range of coloristic effects the artist can achieve, given these restricted means. Ironically, therefore, when it is well done, a "limited-palette" painting should look more or less the same as one created using a "full" palette, as, indeed, Metzger's do. They are very different, though, in both coloring and emotional

*Plate 19*
Eva Borchard
ca. 1943
Oil on canvas
20 × 24 in.

tone from the appealing, but more conventional, portraits made by other contemporary painters, such as the popular Argentine artist Jorge Beristayn.[20] Beristayn, whom Metzger knew in Buenos Aires and who painted her portrait, was celebrated for his images of politicians, diplomats and socialites. He specialized in high colors and bravura brushwork, creating an attractive (but somewhat artificial) atmosphere far removed from Metzger's.

Perhaps the most noteworthy aspect of Metzger's Argentine portraits is the fact that they are neither flattering nor especially cheerful, in the expected sense. Instead, she tends to paint recognizable images of her sitters wearing contemplative or sometimes sad expressions. Her portrait of Eleanor Mallory (fig. 107) is typical. Rather than smiling at the viewer or wearing a neutral expression on her face, Mallory appears to be holding back some strong, and presumably distressing, emotion. Likewise, Lucy Burrows, an American friend who sat for her portrait in 1947, seems ready to challenge something or someone (plate 16). This impression is conveyed partially by her arched eyebrows and proudly held chin. In both cases, as in the earlier portrait of Ambassador Caffery, Metzger places her subjects slightly off-center, with one shoulder higher than the other and the gaze directed at a slight angle. This subtle asymmetry, along with her tendency to paint her sitters against dark, monochrome backgrounds, may again refer back to the 17th-century Dutch portraits with which she was so familiar. Due to the high contrast between dark background and well-lit face, and the lack of a landscape or any extraneous details which might otherwise become distracting, such a format requires the viewer to focus on the sitter's face. Downplaying the background is also a practical decision—a simple, unmodulated, dark background decreases the amount of time that a painting requires.

Burrows gives an interesting account of the painting process. Evelyn Metzger had offered to paint this picture as a going-away gift for Burrows, whose husband had just been assigned to a diplomatic post in the Dominican Republic.[21] At this time Burrows was already in the middle of packing her household effects, her husband was in the hospital and their new baby was ill. Since neither woman had much time to spare, it was lucky that Metzger paints quickly. Burrows estimates that she took the train out to the Metzgers' suburban home only three or four times, sitting for about an hour each session. Though the painting was still wet when Burrows had to pack it, her image arrived in the Dominican capital, Santo Domingo, unscathed.

Aside from these formal portraits, Metzger also painted quite a number of more intimate images of friends and family members in Buenos Aires. One of her favorite sitters was Suzette Telenga Ellinger, whose portrait she painted about twenty times. "Suzie," as both Metzger and Lucy Burrows always call her, also attended Vassar. She first met Evelyn Metzger in New York, however, when she was married to Ham's friend, the diplomat Alan Dawson. In Buenos Aires she lived a few blocks from the Metzgers.

Metzger recalls that her friend always was willing to pose for hours at a time. As Burrows explains, "Suzie was paintable—she had a most interesting face, and spirit." More than that, she seems to have exerted a strange and powerful hold over many Buenos Aires residents.[22] Burrows adds that, even though she disapproved of Suzie's lifestyle and had sworn never to enter her house, "We were all fascinated by Suzie. She kept inviting us, so finally we went, and, once she got you there, she had you." This curious fascination continues. "Still today," says Burrows, "when I get together with my friends from Argentina, we all talk about Suzie."

One does not really get a sense of Suzie's magnetic personality from the rather low-key portraits of a modestly attired, attractive woman reproduced here. The dramatic photo of her standing before her magnificent home (fig. 83) is more impressive in this regard. Still, the paintings are intriguing: a profile view against a mottled gray background, Suzie's hair pulled up, with a single wisp escaping at the neck (plate 8); an evocative portrait of Suzie (accompanied by her own reflection in a gold-framed mirror) reading at a desk, her face half lit and half in shadow, her mouth set in a position of concentration (plate 9). The Suzie series, like the other Argentine portraits discussed here, reveals an important paradox in Metzger's art. On the one hand, she steadfastly avoids direct references to any of the distressing events taking place in Argentina during the ascendancy of Juan Perón, or to current events of the world, such as the effects of World War II.[23] Of course, this is perfectly consistent with the manner in which an Impressionist, Fauve or Cubist artist—or most painters specializing in realistic portraiture— would be likely to act. Then again, rather than emphasizing the attractive, cheerful aspects of her subjects, Metzger imbues many of them with an unexpected sense of introspection, or even melancholy.

Her portraits of her children from this period reveal surprisingly little about their personalities. Metzger's eldest, James, a serious student even as a youth, is seen reading. He is wearing a white sweater which is so loosely painted that it threatens to dissolve into a series of autonomous shapes (plate 18). The double portrait of her sons—a conventional half-length image equally as painterly as the other canvas—stresses the sweetness of the boys' faces—James's more quizzical,

## Experiments in Pure Abstraction

Up to this point in her career, Metzger had concentrated on painting recognizable images in oil on canvas, the most popular artistic method in the Western world for some six hundred years.[28] The 20th century has witnessed a series of radical experiments in art, both in Europe and the United States, however—painting purely abstract images; using nontraditional supports; painting with synthetic pigments and other materials applied in unconventional ways; the exploration of collage. Beginning in the late 1950s, Metzger explored them all.

Partly in an effort to solve the problem of insufficient storage space, Metzger decided to switch from painting on stretched and primed canvases—which take up a great deal of room—to using masonite boards and, later, mahogany plywood panels—which are much thinner and easier to store. Next, in a radical departure from her earlier work, she abandoned both oil paint and the artist's brush, choosing instead to pour liquid (Duco) enamel paint onto each board as she held it flat. She would then move the board, spreading the pigment around with each wiggle and tilt of its support. Naturally, Metzger had very little control over the colored shapes she made—that was the point. In contrast to the art she had created up to this stage, carefully painted from preliminary sketches, these "poured-enamel" paintings could not be planned in advance.

Such experiments obviously have a great deal in common with the work of the Abstract Expressionists, a controversial group of American painters, the most famous of whom was Jackson Pollock. Based in and around New York, these artists came to prominence during the late 1940s and 1950s. Like their work, Metzger's enamel pieces from this period were produced "automatically"—without pre-planning or conscious thought. Also like theirs, Metzger's works are completely abstract, not intended to represent anything from the real world. Yet somehow, again like the Abstract Expressionists', Metzger's art still triggers very personal, almost visceral, responses in the viewer.

Given the fact that Metzger had always been intrigued more by Old Master than avant-garde art[29] and that she had been living in South America throughout the early years of the Abstract Expressionist movement, it is particularly noteworthy that she would make such a break with her previous work in her forties, after decades of success as a representational painter. Nevertheless, Metzger's poured paintings—dramatic abstractions, in which colors spill abruptly into each other, suggesting waterfalls or lava flows—evoke many of the same sorts of responses as Jackson Pollock's "drip" paintings or Franz Kline's outsized slashes of black and

Plate 20
*Eruption*
1959
Enamel on panel
32 × 24 in.

*Plate 21*
*Molten Tide*
1958
Enamel on panel
24 × 32 in.

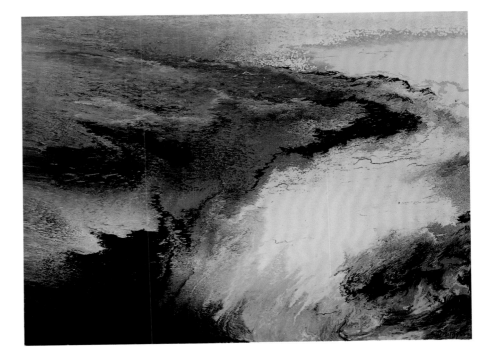

*Plate 22*
*Black, Yellow and Grey*
1959
Enamel on panel
32 × 24 in.

*110*
*Flight into Egypt*
n.d.
Oil on panel
32 × 24 in.

*Plate 23*
Madonna and Child Embracing
ca. 1958
Oil on panel
36 × 24 in.

*Plate 24*
Madonna and Child Enthroned
n.d.
Oil on panel
47½ × 23½

white. Even the titles of these works—such as *Eruption* and *Molten Tide*[30]—sound like entries in an Abstract Expressionist exhibition catalogue (plates 20 and 21).

As Metzger explains, "I was fascinated by Pollock, and made some of my own [dripped pictures]." While she soon found that technique limiting, Metzger went on to produce quite a number of poured works. She says she began using this technique because she wanted to try something radically different, and discovered it was "great fun—[because] you never knew what you would get." Eventually, Metzger tired of this approach and went on to explore additional new directions.[31]

## Success of the 1960s

The 1960s marked a period of extraordinary activity—and success—for Metzger. During the course of this decade, she had no fewer than thirteen solo exhibitions[32]; received an unprecedented amount of critical attention; began to have her work acquired by museums throughout the United States; took five major trips—to Europe, India, Africa and the South Pacific; and continued to try out new themes and techniques.

With her children in their twenties and her husband relatively free from daily business pressures, Metzger blossomed as an artist. One way in which she broadened her aesthetic horizons was through subject matter. Whereas, earlier, she had painted portraits and landscapes more or less exclusively, during the 1960s Metzger made a number of religious pictures, as well. Three examples of this type demonstrate Metzger's interest in traditional Christian iconography—*Madonna and Child Enthroned, Madonna and Child Embracing* and *Flight into Egypt*—all standard subjects, each treated in a different technique.

---

*The rich, glowing quality of these colors, combined with the heavy black outlines separating the principal forms from each other, clearly is intended to evoke the appearance of a stained-glass window, which it does with remarkable success.*

*Madonna and Child Enthroned* (plate 24), a narrow, vertically-oriented painting, is based on a long line of frontal, iconic compositions produced in great numbers from the dawn of Christianity through the 18th century. Metzger was intimately familiar with such images, having grown up with Italian Renaissance Madonnas on the walls at home and having studied this aesthetic tradition in depth during her years at Vassar.[33] Interestingly, these subjects do not reflect her personal religious philosophy. Metzger describes herself as absorbed by the spirituality and love conveyed by religious subjects—the Christian theme of the Madonna and Child, for example—rather than subscribing to any organized religion or dogma. In fact, Metzger seems to have painted considerably more images of religious figures from other, non-Christian traditions, especially those seen on her travels to India and Southeast Asia, than she ever did of Catholic subjects. The closest Metzger comes to explicitly defining her attitude about such spiritual matters occurs in the following quotation from 1993: "To me, art is a kind of religion, the highest creative expression of humanity. It's the only kind of worship I go in for, really."

One especially intriguing feature of the *Madonna and Child Enthroned* is the figures' anonymity and their surprising colors and textures. Metzger's Madonna and Child are highly schematized, with few details of figure or facial expression.

*Plate 25*
Cannes
ca. 1964
Oil and ink on masonite
24 × 32 in.

*Plate 26*
Strasbourg
ca. 1964
Oil and ink on masonite
36 × 24 in.

*Plate 27*
Anemones I
1965
Oil on panel
32 × 24 in.

*Plate 28*
A Difficult Choice
1966
Oil on panel
32 × 24 in.

*Plate 29*
Head of African Woman
1969
Oil on panel
24 × 20 in.

*Plate 30*
Ethiopian Man
1969
Oil on panel
24 × 18 in.

Plate 31
*Washerwomen*
1966
Oil on panel
32½ × 43 in.

*Plate 32*
*Yellow and Blue Bouquet*
1968
Oil on panel
29 × 24 in.

Indeed, each figure has only one eye—the right—and, therefore, lacks the humanizing, individualized quality so common in related paintings by such 15th-century Italian masters as Raphael. Reflecting a more modern sensibility, Metzger has eliminated any reference to angels, seraphim or other attendants. Her colors—deep blues and greens, plus orange and purple—are laid on in thick patches that are unrelated to the realities of anatomy, drapery or three-dimensional space. The rich, glowing quality of these colors, combined with the heavy black outlines separating the principal forms from each other, clearly is intended to evoke the appearance of a stained-glass window, which it does with remarkable success. It may also reflect the artist's long-standing interest in ancient mosaics, which consist of individual, clearly separated patches of color. The Neo-Impressionist canvases of Georges Seurat and Jane Peterson, which Metzger admires, also reflect a similar interest in this early art form.

Unlike the emotionally remote, passive, immobile figures in *Madonna and Child Enthroned,* Metzger's *Madonna and Child Embracing* (plate 23) stresses feelings and physical interaction between the two figures. In this composition, reminiscent of Raphael's often-quoted *Madonna della Sedia,* the Madonna supports her infant's weight with her left hand, caressing his shoulder with her right. The two faces, simplified and overlaid with a series of cubistic diagonal lines, touch, with their foreheads pressing together and halos interlocking. As in the preceding example, large segments of both figures are surrounded by thick black lines, which act as visual "leads" to hold the composition together. Metzger has combined a number of different textural effects, most noticeably, patches of turquoise paint slathered across the Child's leg and Mary's hand with a palette knife, atop areas of heavily diluted brown pigment. Also, especially within the drapery itself, she has created a surprising effect of multicolored patches overlaid with long, narrow ribbons of translucent yellow paint, which bubble and interweave like strands of floating seaweed. To achieve this effect, which Metzger calls "Pollock-ing" because of its spontaneous, uncontrollable qualities, the artist coats varnish over several layers of paint. Then, while the surface is still moist, she pours diluted "strings" of enamel paint over the varnish, causing the yellow pigment to spread out into these strange, organic-looking formations.

Metzger's *Flight into Egypt* (fig. 110) owes less to what she calls her "funny tricks" with textures and more to Cubism, which she had explored in several earlier pictures. Here, however, she has done a far more thorough job of fragmenting the forms of the donkey, the three figures and the surroundings, so that it is sometimes difficult to tell precisely what one is seeing.

Besides these "stained-glass" paintings—so designated because of their technique, rather than their subject matter—Metzger also did a great deal of experimenting

Plate 33
*Green Apples and Pears I*
1969
Oil on panel
24 × 32 in.

with spray-painting in the 1960s. After decades of applying her paint primarily with brushes and palette knives, or by pouring it directly onto the canvas or board, at this time Metzger wanted to try a different approach that involved no physical contact with the painted surface. During the 1970s a whole group of Photo-Realist painters would achieve widespread fame with their use of the airbrush, an industrial tool adopted by fine artists eager to avoid precisely the kind of personal brushwork prized by the Abstract Expressionists. In the preceding decade, Metzger—who had tried the airbrush and determined that it was not for her—opted, instead, for working with humble cans of spray paint. Metzger used this spray technique, alone or in combination with other methods, for landscapes, figure studies and another new subject type—still lifes—with which she achieved considerable critical success.

In works such as *Cannes* (plate 25), the artist begins by spraying in large areas of her composition—the sky, the mountains, the sea, the road. Then, she works over the picture with a narrow brush loaded with thinned black paint. In this manner she adds fluid, linear accents, creating a light and airy tone more reminiscent of a painting by Raoul Dufy than one of her own solid, "stained-glass" works. Sometimes, as in *Cannes,* Metzger also includes sections of drier, thicker paint,

*Plate 35*
Monumental Angkor Figures
ca. 1960
Oil and enamel on panel
32 × 24 in.

*Plate 36*
Angkor Roots (Temple)
ca. 1960
Oil on masonite
32 × 24 in.

Like so many of Metzger's other figurative works, these paintings depict generic, rather than individual, Japanese women—since none has any facial features.

Another group of travel paintings derives from Metzger's visit to India in 1965. Traditional Indian sculpture, mostly religious in nature, is noted for its monumentality and sensuality, two qualities generally lacking in Western devotional art. It is difficult to get a clear sense of scale from Metzger's Indian paintings—since they generally present their subjects from nearby. Even so, her renderings of these massive stone carvings, including the celebrated couple from the façade of the Great Chaitya Hall, a Buddhist devotional space created during the 2nd century A.D. at Karli, in west-central India, *feel* monumental (fig. 115). Metzger's other Indian subjects, which look more like photographic details, clearly display the well-rounded figures, ecstatically closed eyes, and graceful gestures typically associated with such art.

Some of Metzger's most vivid memories of her foreign travels involve the trip she took to Cambodia in 1959. Like countless other visitors to this dramatic site, Metzger was profoundly moved by Angkor Wat, nestled in a jungle in the northwestern part of the country. Built in the early 12th century A.D. under the Khmer king Suryavarman, the huge temple complex is dedicated to the god Vishnu. The temple is protected by high walls, which are themselves surrounded by a broad moat. The enormous stone buildings within the precinct are covered with monumental reliefs and freestanding sculptures representing scenes from the great Hindu epic the *Ramayana*. All aspects of Cambodian culture, including the visual arts, were influenced strongly by India, as can be seen from the images depicted in Metzger's oil paintings. Of the three days she spent at Angkor Wat, the artist especially recalls "that fabulous forest background, these extraordinary roots that would reach sometimes for blocks. They went in and out of the colossal architecture, some chunks [of stone] maybe twenty-feet tall, held up by the trees and roots. . . . The parts that hadn't been restored were the most exciting—sometimes it's hard to tell the roots . . . from the sculpture and architecture. I've never seen more beautiful art."

Several of Metzger's Angkor Wat paintings emphasize the apparent incongruity of the site itself—with acres of masterful carvings largely hidden away among the trees (plate 36). Others show close-up views of specific sculptures, using thick, glowing patches of gem-like blues and greens, surrounded with her signature black outlines (plate 35). This effect is reminiscent of the work of Georges Rouault, a modern French painter noted for his highly expressive, richly colored figurative images which, like Metzger's, have often been compared to stained glass.

As the decade drew to a close, Stuart, her brother, with whom she had always been extremely close and with whom she had undertaken many of her foreign journeys, passed away. As she has always done, no matter what the obstacles, Metzger continued both traveling and working. In 1969 she visited seven African nations and had her fourteenth solo exhibition.[39]

## An End and a New Beginning

Between 1970 and 1974 Metzger had two solo exhibitions and took several trips with Ham—to Portugal, England, France, Scandinavia, Russia, Yugoslavia, Jamaica and the ancient Mayan cities on Mexico's Yucatán peninsula. They also visited major archaeological sites in Turkey with good friends, where Metzger was impressed particularly by the eerie landscapes of the Goreme Valley, the gateway to the mountainous portion of central Turkey known as Cappadocia. This region is famous for its many chapels cut from the soft local rock by the Christian monks who settled there beginning in the 5th century, and for the elaborate Byzantine frescoes that decorate the chapels' interior walls. It is also known for its unusual rock formations—several of which Metzger painted.

Ham's death in 1974 was a major blow. Suddenly, in her early sixties, after forty years of marriage, she was alone for the first time in her life. It took several years for Metzger, determined to rekindle her enthusiasm for living, to again venture overseas. In 1978 she took an "art treasures tour" of Thailand, Hong Kong, Bali, Java and Singapore. She followed this with an even more exotic itinerary—Metzger and her son Edward journeyed to China in 1979, then a far less common destination for Westerners than it has since become. These trips were important to Metzger, not only because of the myriad sights she was able to observe, which provided additional material for her solo exhibitions of 1983 and 1988, but also for their personal, symbolic value. Although she could no longer travel with Ham, she could still travel, and continue to enjoy one of the greatest pleasures in her life.

Following the journey to China, Metzger made one major trip—to the Himalayas in 1980, where she and Edward visited Nepal, Bhutan, Sikkim and Kashmir. Since then, her interest in painting has taken precedence over travel to distant lands. Staying close to home, Metzger has traveled at her easel, producing a series of paintings that capture the essence of three exotic and very different parts of the world—New Guinea, Easter Island and Antarctica—based upon her son Edward's descriptions and photographs collected there in the summer and early fall of 1992. She has also created a multitude of family portraits and landscapes of locations near her Manhattan apartment and her summer home on Long Island. Metzger's output has actually increased during the 1980s and 1990s, in spite of health problems. In fact, this has probably been her most prolific period, with the possible exception of the late 1950s.

# Out of the Stone Age: Images of Papua New Guinea

Set in the Pacific Ocean just north of Australia, the island of New Guinea is an extraordinarily rich and varied place, where more than seven hundred languages are spoken. The eastern half of this island, the independent state of Papua New Guinea, is inhabited mainly by tribes, each of which has preserved its own traditional culture dating back to the Stone Age.

Several of the paintings in Metzger's New Guinea series depict characteristic subjects seen along the banks of the Sepik, New Guinea's widest and longest river—a village woodcarver and a dugout canoe in front of a thatch-roofed house (plate 38), for instance. The majority of the scenes Metzger painted, however, are from the Papua New Guinea highlands, a series of densely populated, but isolated, valleys in the central part of the country which had no contact with outsiders until the 1950s. Until recently, local people hunted and farmed the region, using only bamboo and stone tools. They still live mainly on sweet potatoes and pigs, raising coffee as a cash crop.

Each tribal group in the highlands maintains its own distinctive form of personal adornment, including spectacular combinations of body-painting, wigs, feather headdresses and shell jewelry. As Edward Metzger discovered, the best place for a visitor to revel in the myriad colors, textures and patterns of traditional Papua New Guinea dress is at the annual "Eastern Highlands show." An amalgam of smaller regional celebrations called *sing-sings,* this festival, held in August, brings together thousands of tribespeople from all over the country to spend days dancing, feasting and competing with each other for acclaim.[40]

Several of Metzger's New Guinea paintings are close-ups of typical Huli "wigmen." The male members of the Huli tribe from the Tari River basin are justly

*Plate 37*
Opposite:
*Tribesman with Umbrella*
1993
Oil on panel
32 × 24 in.

*Plate 38*
*Spirit of the Crocodile (New Guinea)*
1993
Oil on panel
24 × 32 in.

*Plate 39*
*Status Symbol (New Guinea)*
1994
Oil on panel
16 × 16 in.

Plate 40
*Highlands New Guinea Woman with Drum*
1993
Oil on panel
32 × 24 in.

*Plate 41*
Back View of Three
New Guinea Women
1994
Oil on panel
32 × 24 in.

Plate 42
*Storm Brewing on Black Lakes, New Guinea*
1994
Oil on panel
24 × 32 in.

famous for their elaborate, hat-like "wigs." Fashioned from human hair, these wigs feature borders of brightly colored blossoms—especially dried yellow daisies—plus extravagant arrangements of feathers. Feathers of choice include multicolored bird of paradise plumes, along with black, white and red feathers from lorikeets, cassowaries and other indigenous birds. Huli men often wear curving boars' tusks, or long thin quills, threaded through their nostrils, as well as thin bands made from blue, yellow and black-speckled snakeskins on their foreheads. In Metzger's pictures we also can see the use of face painting (with pigments made from natural earth, mixed with oil or water, and additional hues derived from commercially available paint); multiple necklaces (made of seeds, quills, shells and other natural substances); net bags, slung over the shoulder and used for carrying personal possessions; and the traditional skirt-like "apron" worn over the genital area, with fresh leaves stuck into the belt to cover the buttocks. In one painting, a white-bearded man leans on what is clearly a non-traditional element—a striped umbrella, one of the most popular trade store items, now regularly used as foul-weather protection in place of the old-style banana leaf (plate 37). Another picture by Metzger illustrates additional cultural artifacts, carried by a woman in this case—the decorative bands often worn on the wrists and upper arms, an impressive crescent-shaped neckpiece cut from the popular gold-lip pearl oyster shell and the long thin, drum seen, and heard, in many tribal settings (plate 40). She also has painted several scenes of tribesmen

*Plate 43*
*Black and White
Highlanders in Procession*
1993
Oil on panel
24 × 32 in.

*Plate 44*
*Mourning (New Guinea)*
1993
Oil on panel
24 × 24 in.

*Plate 45*
*Sunset on the Sepik, New Guinea*
1994
Oil on panel
24 × 32 in.

*Plate 46*
Mud Man,
Burnt Orange Ground
1993
Oil on panel
32 × 24 in.

*Plate 47*
Huli in Headdress, New Guinea
1994
Oil on panel
24 × 24 in.

*Plate 48*
Tribal Leader, New Guinea
1993
Oil on panel
32 × 24 in.

*Plate 49*
*Hulis in Transit*
1993
Oil on panel
32 × 48 in.

who have covered portions of their bodies in white pigment (plate 43), engaging in traditional rituals. One compelling image from the highlands, *Hulis in Transit* (plate 49), captures the startling clash between Stone Age and 20th-century cultures. Here we view not only the literal transportation of a group of workers and family members by truck, but a figurative transit as well—the transition of an entire culture from the way things were (evidenced by the nudity and headdresses of some of the men) to the way things are becoming (demonstrated by the hats, shirts and windbreakers on other men).

The most distinctive of all the highland peoples are the so-called "mudmen" of the Asaro River area (plate 46). Legend has it that the mudmen's ancestors developed these extraordinary helmet-masks (made of gray clay from local streambeds, packed over a framework of woven leaves and other organic materials) to frighten away their enemies. Wearing these masks—sometimes decorated with pigs' teeth or quills—with their entire bodies smeared with mud and pointed bamboo "extensions" on their fingers, the mudmen do appear terrifying.[41]

Metzger's New Guinea paintings are composed in a straightforward manner. Her pictures of the individual Huli wigmen, for example, place each subject on the surface of the painting, very close to the viewer. Even here, Metzger indulges her tendency to separate the subject from the observer psychologically. Just as the models in her academic figure studies look down at the floor or the man on the bench in San Isidro stares into the distance, many of the New Guinea tribespeople depicted by Metzger avert their eyes, bow their heads or position themselves in such a way as to minimize contact. By contrast, the "mudman" confronts the viewer straight on, although his elaborate mask prevents us from having any sense of his emotions or even his looks.

## *Antarctica Snowfields: The Earth's Most Remote Continent*

Just a few weeks after visiting the Eastern Highlands show, Edward Metzger experienced an environment as different as it could be from Papua New Guinea's lush tropical forests—Antarctica. Nearly 5,500,000 square miles of ice and frozen rock, this remote continent seems singularly inhospitable to the human animal. Despite its forbidding aspect, Antarctica does support a few hardy plants (mainly lichens and mosses), an abundance of sea mammals and fish plus some seventy-five species of birds.

*Plate 50*
*Antarctica (Ten Penguins)*
1994
Oil on panel
23 × 32½ in.

*Plate 51*
Opposite:
*Ed in Antarctica*
1993
Oil on panel
24 × 32 in.

*Plate 52*
*Distance (Antarctica)*
1993
Oil on panel
24 × 32 in.

*Plate 53*
Spirit of the Moai
over Easter Island
1994
Oil on panel
32 × 24 in.

*Named to commemorate the day on which it was encountered by the Dutch in 1722, Easter Island sits in the Pacific Ocean 2,300 miles off the coast of Chile. Some six-hundred stone figures or* moai *dot its forty-five square miles. Carved by the island's original Polynesian inhabitants from quarries within the volcano Rano Raraku, the figures are mostly ten to twelve feet tall. A few moai range up to thirty-seven feet and weigh as much as fifty tons. Chile annexed this colorful and mysterious island in 1888.*

*Plate 54*
*Moai in Quarry
(Easter Island)*
1994
Oil on panel
24 × 32 in.

Each summer hundreds of scientists from twenty-four countries go there to conduct ecological, seismological, astronomical and many other forms of research.[42] In recent years Antarctica also has become a popular destination for the more adventure-oriented tour groups, attracted by the romantic isolation of the place, the challenge of dealing with its rugged climate and the chance to approach creatures most people only encounter in zoos.

Metzger's paintings of Antarctica stress several aspects of this remarkable place. One picture focuses on Edward, dressed in a bulky red parka and knitted hat, kneeling on the ice and grinning delightedly as a group of penguins interact all around him (plate 51). Several others point up the friendly, often humorous, qualities of these creatures, with which humans seem to have a strong affinity (plate 50). At the same time, its views of barren snowfields and unsettled sky create an eerie quality. A third painting shows a visitor walking away, toward a group of four tiny, red-clad figures immeasurably far back in the frozen distance (plate 52). This last example, with its emphasis on the utter vulnerability of the human species, has some of the same emotional qualities as Metzger's more contemplative scenes of people in Central Park.

*Plate 55*
Evanier Window
1991
Oil on panel
48 × 32 in.

*Plate 56*
Winter View of Evanier
from Birches
1992
Oil on panel
24 × 32 in.

# Landscapes and Family Portraits

Metzger has never stopped experimenting. In 1990, unwilling to discard several beautiful, but threadbare, hand-embroidered pieces of 16th-century Italian silk, Metzger hit upon the idea of using them in collages. So she did, cutting around the embroidered sections, attaching them to wood panels and painting around them, using the fabric segments for additional textural effects. Continuing experimentation aside, many of Metzger's paintings since 1980 have been portraits of her grandchildren, along with evocative views—which, in a sense, one could also call "portraits"—of her daughter Eva's new Denver home, her own vacation house on Long Island and Metzger's favorite people-watching site, Central Park.

*Plate 57*
*Chandra in the Snow*
1993
Oil on panel
32 × 24 in.

*Plate 58*
Opposite:
*Eva in Evanier*
1994
Oil on panel
32 × 24 in.

*Plate 59*
*Eva Looking over Her Shoulder*
ca. 1962
Oil on canvas
15½ × 11½ in.

Chandra—Eva's daughter from her first marriage—appears often in Metzger's work. In one particularly attractive example, Eva leans over a changing table to gaze at her infant daughter. The obvious emotional connection between mother and child, and the unusual double perspective provided by their mirror image, calls to mind pictures by Mary Cassatt (plate 60). Several years later, her proud grandmother painted Chandra in a lively close-up, the brushstrokes in her blue blouse positively vibrating with energy. Chandra also appears in a number of genre situations—showing off a pretty robe-and-nightgown set (plate 61) or walking through the snow in her garden, moving her hands with a quizzical look on her face, as though in the middle of a long story (plate 57).

The informality of these images contrasts markedly with the *Self-Portrait with Chandra* (page 118) a rare subject for this artist, who maintains that she dislikes posing, for herself or anyone else. Here, the figures' elegant apparel, poses and expressions all indicate that some special occasion is at hand.

Metzger has also painted several studies of Eva's Colorado home—a large, unusual and extremely handsome structure. One view, from the interior of the house, cleverly juxtaposes a colorful bouquet of cut flowers on one side of the picture windows with the wintry landscape on the other (plate 55). Metzger's exterior view of the house, as observed through a stand of trees, has a pleasing, airy feel (plate 56). It is especially interesting to note the dots and dashes of color the artist has used here to signify bits of grass and dead leaves revealed by the melting snow. These shimmering, pointillist passages hark back to the landscapes of Jane Peterson and the Canadian-born Neo-Impressionist painter Maurice Prendergast. Moreover, the way in which small pieces of color are treated as distinct entities, totally separate from the rest of the picture, again may reflect Metzger's interest in mosaics.

Metzger's pictures of James and Agnes's son, Jimmy, feature more than a touch of Christopher Robin-like winsomeness—as the little boy sits pensively under the family Christmas tree, clutching a large, floppy stuffed animal (plate 66) or poses outdoors in his bathing suit, the summer sunlight streaming through the slats of the fence and glancing off his shoulders and tousled hair (plate 65). There are also quite a number of fetching portraits of Elizabeth, James and Agnes's younger child. Metzger has caught Elizabeth in a number of different situations—playing

*Plate 60*
*Eva and Chandra (Mirror)*
1983
Oil on panel
23 × 31¼ in.

*Plate 61*
*Chandra en peignoir*
1990
Oil on panel
29 × 21½ in.

*Plate 62*
Opposite:
*Self-Portrait with Art*
1994
Oil on panel
32 × 24 in.

*Plate 63*
Below:
*Elizabeth*
1990
Oil on panel
21½ × 29 in.

*Plate 64*
*Elizabeth Modeling Yellow Hat*
1991
Oil on panel
15¼ × 18½

*Plate 65*
Below:
*Jimmy at Quogue Pool Fence*
1985
Oil on panel
32 × 24 in.

*Plate 66*
*Jimmy at Three
with Teddy Bear*
1985
Oil on panel
40 × 30 in.

*Plate 67*
At Water's Edge, Quogue Beach
1986
Oil on panel
24½ × 32 in.

*Plate 68*
Boy Architects on Quogue Beach
1986
Oil on panel
24 × 32 in.

*Plate 69*
*Tasseled Grass, Quogue Refuge*
1987
Oil on panel
24 × 31 in.

*Plate 70*
*White Crescent, Turquoise Water*
1980
Oil on panel
32 × 48 in.

*Plate 71*
*Anemones in Greece*
1978
Oil on panel
32 × 48 in.

with twigs (perhaps initiating her own form of artistic exploration, plate 63) and thoughtfully fingering an enormous straw hat, with stylized images of the ocean and seabirds arranged in a repeating decorative pattern, almost like wallpaper, behind her (plate 64).

While most of these images are sentimental, attractive and emotionally neutral, nevertheless, Metzger's young subjects do seem to be generally quiet, even subdued. The artist achieves a quite different effect painting figure studies of children playing on the beach near her summer home in Quogue, on the eastern end of Long Island. Here, Metzger uses an Impressionist approach, freezing the action of the boys digging in the sand (plate 68) and the little girls frolicking at the water's edge (plate 67). Impressionist as well are the light palette and the huge diagonal expanse of empty sand she has left in the lower right corner of plate 67. In her characteristic way, Metzger has virtually hidden the children's identities; while their body language remains expressive, the artist has turned her subjects' faces away from us or hidden them in deep shadows.

Although there are relatively few portraits of James or Edward as grown men, Metzger recently has painted several pictures of a Thai friend of Edward's named Mallika. She appears at the very bottom of one striking composition, her mouth set in a serious expression, eyes masked by dark glasses, head tilted up (plate 73). In another she sits in a bright red dress, smiling broadly beneath an

Plate 72
*Mallika in Red*
1993
Oil on panel
32 × 24 in.

*Plate 73*
*Mallika in Costa Rica*
1993
Oil on panel
32 × 24 in.

astonishingly active, abstract pattern which might represent another painting or a fabric wall-hanging (plate 72). At first the viewer's eye naturally focuses on Mallika, but it is difficult to resist looking at that pattern. With its bold swirls of purple, beige and blue arranged in irregular groups of dotted lines, it has a powerful and distracting effect similar to that created by the background of several of van Gogh's portraits, most notably the images of M. and Mme. Roulin.

Like so many of the subjects she paints with regularity, the Quogue house—and the landscape surrounding it—are intimately familiar to Metzger, since she has been going there for nearly forty years.[43] She has painted the local grasses, trees and water, in handsome compositions, with feathery shapes constructed of creamy strokes of paint, during every season and at various times of day (plate 69). Over the past two decades Metzger has reserved the major part of her considerable energies and attention, however, for an extraordinary series of landscapes painted in a very different part of New York—namely, Manhattan's Central Park.

## The Central Park Series

Evelyn Metzger has spent her entire life in close proximity to urban parks. Her childhood home adjoined Riverside Park, that long strip of Upper West Side greenery running along the Hudson River. During the 1940s she spent a good deal of time sketching the many beautiful parks of Buenos Aires. For the past four decades, Central Park has been a virtual fixture in her life. She goes there at every opportunity—her apartment is just two blocks away, and she can see the park from the terrace of her glassed-in rooftop studio.

There are no extant parkscapes by Metzger from her earliest years, and relatively few from Argentina. In contrast, a recent inventory of the paintings stored in her Manhattan apartment revealed several hundred images of Central Park—one of the few parks large enough and varied enough to accommodate even Metzger's obsessive interest. Designed by Frederick Law Olmsted in 1856, the park fills an oblong area of central Manhattan measuring 840 acres between 59th and 110th Streets. Fully 150 of those acres are water—lakes and ponds. The remainder comprises rolling lawns, wooded hillsides, flower gardens, bike and bridle paths, walking trails and sports fields. Special features of the park include the open-air Delacorte Theater, a boat house for renting rowboats and other pleasure craft, a skating rink, playgrounds and a zoo. There is also a great deal of public art in and around Central Park, and a number of major art museums lining the streets that border it (Fifth Avenue and Central Park West). Two of these, the Frick Collection and the Metropolitan Museum of Art, have played an important part in Metzger's life.

No matter what degree of success she may have achieved in the past, an artist has to believe that her current works are among her very best. Otherwise, why go on making art? Fortunately, many people, including this writer and Metzger herself, believe that her Central Park paintings are some of the strongest she has ever produced. In part this is due to Metzger's intimate knowledge of her subject.

As a mature artist who has devoted many years of her life to painting the natural wonders she sees around her in all kinds of weather and times of day, Metzger inevitably invites comparison with the French Impressionist Claude Monet. Indeed, there are many parallels between Metzger's Central Park pictures and Monet's late series—of Rouen Cathedral, haystacks, poplar trees and, especially, water lilies. Monet painted water lilies for some twenty-seven years—under every conceivable circumstance, in innumerable configurations, using wildly varied colors, textures and techniques. He was so devoted to this subject, in fact, that he installed his own custom-designed water garden at his home in Giverny. While Metzger has to walk two long city blocks to reach her equivalent of Monet's garden, her devotion to it is comparable. Like the French master, she has painted her corner of nature in numerous formats—large and small, horizontal and vertical.

*Plate 74*
Opposite:
*Park Bench with Shadows*
1982
Oil on masonite
32 × 24 in.

*Plate 75*
Overleaf:
*San Remo Reflections, Central Park*
1994
Oil on panel
24 × 32 in.

*116*
Edward Hopper
*People in the Sun*
1960
Collection National Museum of American Art, Smithsonian Institution

*Plate 76*
*Tulip Magnolia at the Frick*
1986
Oil on panel
24 × 32 in.

Plate 82
*Dog-Walker*
ca. 1979
Oil on panel
24 × 32 in.

*Like her other subjects, Metzger's Central Park pictures have strong compositions, based on prominent diagonals that recede into the distance and the irregular placement of basic design elements — a large tree, a lamppost, a colorful flower bed.*

*Plate 83*
*Chatting at 59th Street Plaza*
1991
Oil on panel
32 × 48 in.

*Plate 84*
*Baby Watching Musicians, Central Park*
1993
Oil on panel
24 × 32 in.

*Plate 85*
*School Class, Central Park
(Glancing Back)*
1992
Oil on panel
32 × 48 in.

*Plate 86*
*Family and Tricycle at the Pond,
Central Park*
1993
Oil on panel
24 × 32 in.

*Plate 93*
*Sunbathers on Grass,*
*Central Park*
1989
Oil on masonite
32 × 48 in.

*Plate 94*
*Willows, Central Park*
1991
Oil on panel
24 × 32 in.

*Plate 95*
*Sunny Day in Park with Fence*
1993
Oil on panel
24 × 32 in.

Plate 96
*Chess in the Park*
ca. 1982
Oil on masonite
24 × 32 in.

snow that ultimately makes this more than simply another charming winter landscape.

Likewise, Metzger stresses human interest in her pictures of a mounted policeman (and the reactions of other people to his arrival on the scene), and of a couple strolling through the park. The latter are walking side by side, both facing forward but with a sense of communion reflected in their similar outfits and closely-matched strides. Metzger continues her exploration of contemporary urban genre painting in a view of two pairs of men seated at outdoor tables in the park playing chess and cards (plate 96). Even though their faces are largely hidden from us, their body language conveys a sense of intense concentration and camaraderie.

What keeps Metzger's Central Park series from becoming stale is the remarkable variation among the paintings—both the subjects themselves and her handling of them. As Metzger has commented, in the park "there's always something going on. It entertains me [because] there's *always* a different mood." Note, for example, the audaciousness of the painting of children walking by the park on their way to school. One smiling little girl suddenly turns to say something to a person behind her, gesturing with the hand that holds her sack lunch, as the parade of New Yorkers—all ages and shapes—passes by (plate 85). This anecdotal picture

*Plate 97*
*Falling Snow, Central Park*
1985
Oil on panel
32 × 48 in.

235

Plate 98
*Blue Carriage with Brown Horse, Central Park*
1992
Oil on panel
24 × 24 in.

*"I was always fascinated by carriages, in Italy, Buenos Aires, wherever."*

*Plate 99*
Red-Lined Carriage with White
Horse, Central Park
1992
Oil on panel
24 × 32 in.

*Plate 100*
Below:
White Carriage Pulling Away
ca. 1992
Oil on panel
24 × 32 in.

*Plate 101*
*Winter Bridge with Lamppost,*
*Central Park*
n.d.
Oil on panel
24 × 32 in.

*Plate 102*
*Tavern Terrace Panorama, Central Park*
1993
Oil on panel
24 × 32 in.

is quite different from Metzger's sensuous study of sunbathers relaxing under the trees in Central Park. Since the figures in this work are much further away from us, their individual personalities are not revealed. In any case, this painting is less about a story than about warmth and light—as demonstrated most clearly by the reclining figure wearing a bathing suit (a quotation from Matisse or possibly Ingres) on the right side of the work (plate 93).

In yet another vein Metzger has painted several images of the park's Bethesda Fountain, a large circular basin with Emma Stebbins's monumental bronze sculpture *Angel of the Waters* (1868) at its center. Here, the focus is neither stories nor sensuality, but rather the panoramic view, the overall mood and the formal qualities inherent in the scene. Metzger is particularly adept at establishing pleasing visual rhythms in her pictures—seen here, for example, in the patterns of light and dark made by the shadows of trees on the hillside and the low stone wall on the grass and—in the upper right-hand corner—the dramatic contrast lent by a few light-toned branches (plate 88).

Metzger's Bethesda Fountain series also demonstrates one of the recent experiments with which she is most pleased: her use of black underpainting and black borders, along with the touches of black she scatters throughout even her brightest pictures to increase their visual contrast. This technique, which she has been using periodically for the last few years, adds an additional air of richness and excitement to her compositions. In a comment that comes as close as Metzger ever gets to self-praise, she grudgingly says that these black-accented pictures are "not too bad."

Water is a recurring theme in these Central Park scenes. The delicate, shimmering quality of moving water invariably adds a great deal to Metzger's studies of the Central Park boat house and its popular outdoor café. Here, too, she manages to create tremendously varied scenes. One picture emphasizes the chatter and closeness of a large group of patrons, while another focuses on a solitary young woman whose neck is warmed by the sun as she sits alone, reading her newspaper (plate 89). In still another variation on this theme, Metzger has caught an elderly lady—her hair covered by a turban, eyeglasses slipping down her nose—as she sits down to (or, possibly, rises from) her table.

Scale is the key to Metzger's painting of what appears to be a small fleet of sailboats, softly gliding against a backdrop of summer trees. Yet something seems to be amiss as one compares the size of the boats with the scale of the people observing them from the water's edge, until it becomes clear that these are toy boats, remote-controlled models that provide a pleasant respite from the world outside the park.

Scale is also the first thing one notices about *Camels in Central Park,* one of Metzger's most overtly humorous pictures (plate 87). Aside from the essential incongruity of seeing these behemoths grazing amid the skyscrapers of midtown Manhattan, the painting itself is huge—nearly four by eight feet—and, so, doubly hard to miss. Although Metzger had ridden camels in Egypt many decades earlier, their juxtaposition with the New York skyline is startling and purely imaginative. Metzger has also painted vivid images of the Central Park Zoo—focusing on polar bears, seals and other animals.

---

*It is her art which gives meaning and zest to Metzger's life. Whether in town or on Long Island, she still paints every day, going into the studio by 9:00 A.M. and working straight through, with short breaks for meals, until evening.*

*Plate 103*
*Park Bench with Young Child*
1982
Oil on masonite
32 × 26 in.

*Plate 104*
*Homeless Man with Shopping Bag, Central Park*
ca. 1991
Oil on panel
24 × 24 in.

Plate 105
*Melancholy (Central Park)*
ca. 1990
Oil on panel
24 × 32 in.

*Plate 106*
*Schemmy's Ice Cream Parlor, Rhinebeck*
1994
Oil on panel
24 × 32 in.

*Plate 109*
Pas de Deux (Black Costumes)
1994
Oil on panel
24 × 24 in.

Plate 110
*Ruisdael Redux*
1994
Oil on panel
24 × 32 in.

Plate 111
Opposite:
*Brueghel Redux*
1994
Oil on panel
24 × 32 in.

*Terrace Panorama* (plate 92 and 102), in which far more space is devoted to the gaily decorated lamps hanging among the trees than to the summarily executed people and furniture below, with Metzger's youthful depiction of a tea dance at the Royal Poinciana (fig. 12). She would like to see still more changes. With an artistic career that has already spanned three-quarters of a century, moving from Art Deco drawings to Fauve landscapes, expressionist abstractions and many variations on Impressionism, Metzger—like any serious artist—is focused on her future, not her past. "I have a feeling I'm ready for some kind of change." she says, "but I haven't found it yet."

As part of this continuing search, in 1994 Metzger also began experimenting with paintings of ballet dancers (plates 108 and 109), reflecting her lifelong interest in depicting physical movement and—perhaps the most surprising development of all—a series of variations on Old Master paintings collected by the family. The latter have nothing in common with the slavish copies essayed by youthful art students. Instead, they are personal evocations of the spirits of the originals, created by unexpected means. For example, in *Ruisdael Redux* (plate 110) Metzger has reproduced only the principal masses of the 17th-century Dutch landscape-painter's subjects, blocking them in with juicy sets of vertical strokes set against black, to produce a highly stylized effect. Even more remarkable is

"Art requires a great deal of love. You have to love art to create it and to appreciate it. . . . For me, art is a real reason for striving, for wanting to stay alive, for continuing to produce and enjoy."

her reinterpretation of Brueghel's *Wedding Dance.* In Metzger's version, seemingly random patches of color are overlaid—and tied together—by fluid, sketchy, cartoon-like outlines of gold paint in a manner that visually distills the Brueghel original (plate 111).

Evelyn Metzger continues walking through Central Park, contemplating the dune grass at Quogue and painting the people and places she knows best. She observes: "Art requires a great deal of love. You have to love art to create it and to appreciate it. . . . For me, art is a real reason for striving, for wanting to stay alive, for continuing to produce and enjoy."

# Notes

1. Interview of the artist by Brett Topping (8/7/92). Unless otherwise indicated, direct quotes attributed to Evelyn Borchard Metzger are taken from a series of interviews she gave to Ms. Topping and the author, dated: 8/7/92, 12/10/92 and 5/25–26/93, or telephone conversations from 6/18/93, 7/19/93, 7/24/93 and 7/29/93.

2. Karl Lilienfeld (1885–1966) was born in Germany, where he received his Ph.D. in art history. After serving as assistant director of the Royal Art Galleries in The Hague, and then director of the Leipzig Art Museum, he emigrated to the United States in 1915. Here, he started his own gallery on East 57th Street in New York in 1932. Later renamed the Van Diemen-Lilienfeld Galleries, it presented a major exhibition of Metzger's work in 1966.

German-born Wilhelm Reinhold (W.R.) Valentiner (1880–1958) was educated at the universities of Heidelberg and Leipzig. His doctoral dissertation was on the art of Rembrandt. After three years at the Kaiser Friedrich Museum in Berlin, Valentiner came to New York, where he began a distinguished career in the museum world. He was a curator at the Metropolitan Museum of Art; director of the Detroit Institute of Arts; co-director of the Los Angeles County Museum of Art; consultant to the J. Paul Getty Museum; and founding director of the North Carolina Museum of Art in Raleigh. He was also the founder and editor of both *Art in America* and *Art Quarterly*. See Margaret H. Sterne, *The Passionate Eye: The Life of William R. Valentiner* (Detroit: Wayne State University Press, 1980).

Wilhelm von Bode (1845–1929) was a distinguished art historian and the director of the Kaiser Friedrich Museum in Berlin, now the Wilhelm von Bode Museum. There he served as Valentiner's mentor and produced an enormous number of German-language articles and books concerning various aspects of Italian and Dutch Old Master artworks. His best-known volume available in English is *Great Masters of Dutch and Flemish Painting*, trans. Margaret L. Clarke (1911; reprint, Freeport, N.Y.: Books for Libraries Press, 1967).

3. Two of these essays—one on Spanish Romanesque ivories, the other concerning a religious picture by Giovanni Bellini—were published in the *Vassar Journal of Undergraduate Studies* 6 (May 1932).

4. Born in Pelham, New York and trained in Paris by Jean-Léon Gérôme, the noted master of academic painting George Bridgman (1844–1943) instructed thousands of students at the Art Students League. Although considered old fashioned today, the system of drawing he devised—known as "constructive anatomy"—was popular for many years. Its principles were collected and published in a series of six books, beginning in 1920.

5. As Metzger presumably was aware, a number of Old Masters, notably Leonardo da Vinci, made sheets of drawings of grotesque heads.

6. For the young Evelyn Metzger, of course, the "flapper" was not the quaint historical oddity she seems today. A popular symbol of the emancipated female in post-World War I America, with her newly acquired right to vote, her daringly short skirts, bobbed hair and love for speakeasies, the flapper was an exciting kind of woman, whom Metzger often would have seen. She notes that she did go to speakeasies later in the 1920s and remem-

bers being very impressed by the chic dressers she saw at this time in Harlem hot spots, such as the Cotton Club.

7. Today, the Horace Mann High School is a highly regarded, coeducational institution that includes art in its curriculum. It is located where the Boys' School used to be, in the Riverdale section of the Bronx.

8. Evelyn first formally studied French at Horace Mann. As she says, she already had a background in French through her travels, readings and family friends. At Vassar she studied German and Italian, the two languages traditionally required for a degree in art history. Both Spanish and Portuguese she acquired "on the job," when living in Colombia and, later, Brazil.

9. From a telephone interview, 7/19/93.

10. Agnes Millicent Rindge Claflin (1900–1977) was a prominent university teacher, author and museum administrator. She taught art history at Vassar from 1923 until her retirement in 1967. After earning her Ph.D, in art history from Radcliffe College (1928), Rindge received numerous awards and wrote a book entitled *Sculpture* (New York: Payson & Clarke, 1929). She also served as a vice-director of New York's Museum of Modern Art during World War II; thereafter, she periodically acted as an advisor to the museum.

11. Sally James Farnham (1876–1943) began her career at the age of twenty-five, when her husband, a designer of silver for the Tiffany Company, gave her some clay with which to pass the time during a lengthy hospital convalescence. Excited by her initial explorations—and encouraged by Frederic Remington, a friend from her hometown of Ogdensburg, New York—she quickly progressed from modeling images of cowboys and horses to designing fountains, war memorials and portraits of prominent politicians and arts personalities.

12. According to Farnham's son John (quoted in the *New York Times* 16 March 1989, Sec. 3, 13:1), the artist even researched the buttons on General Bolívar's uniform.

13. Jane Peterson (1876–1965) was a painter of vibrant watercolor and oil paintings in a Neo-Impressionist style. Born in Elgin, Illinois, she spent her early years working as an art teacher, to save the money necessary to support her further training—in New York, with Frank Vincent DuMond; in Paris, with Jacques-Emile Blanche; in London, with Sir Frank Brangwyn; and in Madrid, with Joaquín Maler Sorolla y Bastida. She also traveled as much as she could, to such areas as the Middle East, where single foreign women seldom went alone.

Noted for her sunny landscape views of Europe and New England, Peterson was one of the most popular artists in New York during the 1920s. She taught at the Art Students League, and received many awards. After her marriage to a successful corporate lawyer in 1925, Peterson acceded to her husband's wishes and stopped traveling. She concentrated instead on paintings of flowers, often with gold- or silver-leaf backgrounds, which were very well received. On her husband's death in 1929, Peterson inherited a great deal of money, which allowed her to resume her travels. She continued to paint throughout her life. See J. Jonathan Joseph, *Jane Peterson: An American Artist* (Boston: private printing, 1981).

14. Metzger notes that at one point during her second year at Vassar her parents owned eighty 17th-century Dutch paintings.

15. The United States did not yet have an embassy in Colombia. Caffery later became United States ambassador to Brazil, France and Egypt. Among his outstanding protégés were the Metzgers' friends Freeman ("Doc") Matthews, Ellis Briggs and Walter Donnelly, all of whom became ambassadors themselves.

16. Interview with Mrs. Charles (Lucy) Burrows by Brett Topping and the author in Washington, D.C., 6/23/93. All other quotations attributed to Mrs. Burrows also come from this meeting.

17. Late that evening, Metzger recalls, she, Ham and several friends took a singularly romantic carriage ride home from the elegant Casino in the Park.

18. Jean Boley, "Letter from Buenos Aires," *New Yorker,* 16 September 1944, 51. Boley also published several books of fiction. She and her husband, Herman Boley, met the Metzgers in Buenos Aires, where Mr. Boley was head of the International Harvester Company.

19. As with other modern art movements, Metzger was familiar with the Fauve style, which had developed in Paris around 1905. In fact, she met its erstwhile leader, Henri Matisse, at an art exhibition.

20. I would like to thank Mrs. Burrows for bringing the works of this painter to my attention.

21. Evelyn Metzger has given almost all of the portraits she has painted to their subjects as gifts.

22. From stories about her it appears that Suzie had a forceful personality and a great desire to make an impression.

23. Jean Boley points out that everyone in Buenos Aires seemed to ignore the war, "Letter from Buenos Aires," 49. Metzger does recount several stories about having Juan Perón to dinner at their home.

24. Because of the vagueness of the titles listed in the catalogue, it is difficult to be certain precisely which paintings were displayed. Of those discussed here, *Llao Llao Trees, Three Carriages Belgrano R, San Isidro Church* and *Eva with Cat* definitely were on exhibition. The sculptures shown were individual busts of Ham and the children.

25. I would like to thank Ralph Pemberton for his invaluable help in translating this, and other, documents.

26. Metzger says that one of her largest receptions occurred in 1965, at the Norfolk University Museum (now the Chrysler Museum), which some six hundred people attended. The show at the Mexican-American Cultural Institute in 1967 also drew crowds.

27. Of course, Metzger had already experimented with Cubist structure, in one of her Cartagena cityscapes from the late 1930s.

28. She had also experimented, briefly, with monotypes.

29. Metzger was a young art student living in New York City during the 1920s, when several of the most influential avant-garde art institutions began. Marcel Duchamp, Man Ray and Katherine Dreier founded the Société Anonyme in 1920; Alfred Stieglitz's Intimate Gallery started up in 1925; A. E. Gallatin's Gallery of Living Art was established at New York University in 1927; and the Museum of Modern Art opened its doors in 1929. While she was aware of these activities, Metzger was not directly involved in them.

30. Like many other artists, Metzger seldom gives her work titles, until and unless she must for exhibition or publication purposes.

31. Although she used thicker pigment and produced bolder designs, Metzger's poured enamel abstractions seem related to the so-called "stain" paintings of Helen Frankenthaler and Morris Louis, also developed during the 1950s.

32. These include a 1963 retrospective with over fifty paintings held at Vassar College, which was attended by faculty, students, the college president and Agnes Rindge.

33. Metzger frequently refers to a painting in Antwerp, a Madonna and Child by the 15th-century French painter Jean Fouquet, as "one of my very favorite paintings."

34. In 1991 Metzger donated about fifty of these hats to the Brooklyn Museum.

35. As Brett Topping has pointed out, the repeating pattern created by this screen looks rather similar to the decorative pattern seen in the background of several novelty photographs taken of Evelyn and Edward Metzger during a trip to Spain (see page 106).

36. Three reviews by M-L, D'Otrange Mastai, in *The Connoisseur* (London) 153, no. 615 (May 1963): 69; 157, no. 634 (December 1964): 274–75; and 163, no. 658 (December 1966): 279.

37. Aymel Seghers, *The Arts Review* (London) 15, no. 7 (April 20–May 4, 1963): 4. It is noteworthy that Metzger's *Bouquet Printanier* (1963) was selected for the cover of this issue of *The Arts Review* and reproduced in full color.

38. "C.N.," *Arts,* 1966.

39. This 1969 exhibition was held at Albion College, in southern Michigan.

40. According to *National Geographic,* in 1969 the Mount Hagen *sing-sing* attracted 60,000 tribespeople, many of whom walked barefoot for over one-hundred miles to get there. See Robert J. Gordon and David Austen, "Papua New Guinea: Nation in the Making," *National Geographic* 162, no. 2 (August 1982): 142–48, and Malcolm S. Kirk, "Journey into Stone Age New Guinea," *National Geographic* 135, no. 4 (April 1969): 568–92. For additional information and stunning, close-up, photographic portraits of highlands people in full regalia, see Malcolm Kirk's *Man as Art: New Guinea* (New York: Viking, 1981).

41. The German artist Rebecca Horn (b. 1944) has also experimented with "finger extensions" in her sculpture and film work.

42. For details on their day-to-day experiences, see Michael Ryan, "Why They Come to the Ice," *Parade* (July 11, 1993): 4–5.

43. Stuart Borchard purchased this house in the 1950s. The artist was a frequent visitor to Quogue during his lifetime. Since inheriting it from her brother, she has made various improvements to the house, adding a third floor and pool.

44. The image of falling snow may have been borrowed from one of the 19th-century Japanese woodcuts which the Impressionists held so dear, and which Metzger's friends Ambassador and Mrs. William Leonhart collected. For a catalogue of the Leonharts' collection, see Ann Yonemura, *Yokohama: Prints from Nineteenth-Century Japan* (Washington: Arthur M. Sackler Gallery, Smithsonian Institution, 1990).

45. See, for example, D'Otrange Mastai in *The Connoisseur* 163, no. 658 (December 1966): 279.

46. J. Jonathan Joseph, *Jane Peterson: An American Artist* (Boston, private printing, 1981), 106.

# Evelyn Borchard Metzger

### Solo Exhibitions

1950 Galería Muller, Buenos Aires, Argentina
1962 S.A.G. Gallery, New York
1963 Everhart Museum, Scranton, Pennsylvania
1963 Vassar College Art Gallery, Poughkeepsie, New York
1963 Galerie Bellechasse, Paris
1964 Frank Partridge Gallery, New York
1965 Norfolk Museum of Art (Chrysler Museum), Norfolk, Virginia
1965 Gibbes Art Gallery, Carolina Art Association, Charleston, South Carolina
1966 Columbus Museum of Arts and Crafts, Columbus, Georgia
1966 Van Diemen-Lilienfeld Galleries, New York
1966 Georgia Museum of Arts, University of Georgia, Athens, Georgia
1966 Telfair Academy of Arts and Sciences, Savannah, Georgia
1967 Mexican-American Cultural Institute, Mexico City, Mexico
1967 Hudson Museum, University of Maine, Orono, Maine
1967 Slater Memorial Museum and Converse Art Gallery, Norwich, Connecticut
1969 Albion College, Albion, Michigan
1970 Graham-Eckes School, Palm Beach, Florida
1973 Bartholet Gallery, New York
1983 Arsenal Gallery, New York
1988 Quogue Library, Quogue, New York

### Works included in the following collections

Albion College, Albion, Michigan
Lyman Allyn Art Museum, New London, Connecticut
Arizona State Museum, Tucson, Arizona
Art Association of Richmond, Richmond, Indiana
Baldwin-Wallace College, Berea, Kentucky
Butler Institute of American Art, Youngstown, Ohio
Chrysler Museum, Norfolk, Virginia
Denver Art Museum, Denver, Colorado
Evansville Museum of Arts & Science, Evansville, Indiana
Finch College, New York
Robert Hull Fleming Museum, University of Vermont, Burlington, Vermont
Georgia Museum of Art, University of Georgia, Athens, Georgia
Gibbes Museum of Art, Charleston, South Carolina
Grand Rapids Art Museum, Grand Rapids, Michigan

Hudson Museum, University of Maine, Orono, Maine
Frances Lehman Loeb Art Center at Vassar College, Poughkeepsie, New York
Lowe Art Museum, Coral Gables, Florida
Mills College Art Gallery, Oakland, California
Mount Holyoke College Art Museum, South Hadley, Massachusetts
Museum of Art and Archaeology, University of Missouri, Columbia, Missouri
Museum of Contemporary Art, San Diego, California
Philbrook Museum of Art, Tulsa, Oklahoma
San Diego Museum of Art, San Diego, California
Sheldon Swope Art Museum, Terre Haute, Indiana
University Gallery, University of Massachusetts, Amherst, Massachusetts
Washington County Museum of Fine Arts, Hagerstown, Maryland
Western New England College, Springfield, Massachusetts
Also represented in the Art in Embassies Program of the United States Department of State.

## Bibliography

*Noticias Gráficas* (Buenos Aires), August 30, 1950
*La Nación* (Buenos Aires), September 5, 1950
*Art Voices,* December 1962
*New York Herald Tribune,* December 8, 1962
*France-Amérique,* December 25, 1962
*Arts* (Paris), January 12, 1963
*Vassar Alumnae Magazine,* April 1963
*Arts Review* (London), April 1963
*The Connoisseur,* May 1963
*New York Herald Tribune* (European edition), December 4, 1963
*Nouveaux Jours* (Paris), December 13, 1963
*The Connoisseur,* December 1964
*Norfolk Museum of Art Catalog,* February 28, 1965
*The Connoisseur,* December 1966
*The News* (Mexico City), February 2, 1967
*El Sol* (Mexico City), February 5, 1967

*Dictionary of International Biography*
*International Who's Who in Art and Antiques*
*International Who's Who of Professional and Business Women*
*New York Art Review*
*Women of Achievement*
*Who's Who in American Art*
*Who's Who of American Women*
*Who's Who in the East*
*World Who's Who of Women*

# List of Art Reproductions

## Figures

Figure 12:
*Coconut Grove Tea Dance*
ca. 1921

Figure 14:
*Deco Style*
ca. 1921

Figure 15:
*Triangle Lady*
ca. 1921

Figure 16:
*Young Sophisticate*
ca. 1921

Figure 96:
*Fantasy Man*
ca. 1921

Figure 97:
*Inkwell Nude*
ca. 1921

Figure 101:
*Bogotá Resident*
ca. 1935

Figure 102:
*Bogotá Resident*
ca. 1935

Figure 103:
*Bogotá Resident*
ca. 1935

Figure 104:
*Cartagena, Colombia ("The Tanker")*
ca. 1934-35

Figure 105:
*Cartagena, Colombia*
ca. 1957

Figure 107:
*Eleanor Mallory*
ca. 1945

Figure 109:
*Eva Borchard at Four*
1952

Figure 110:
*Flight into Egypt*
n.d.

Figure 111:
*Cave Mural*
n.d.

Figure 112:
*Making Up*
ca. 1960

Figure 113:
*Geishas*
1961

Figure 114:
*Khmer Idol*
ca. 1960

Figure 115:
*Indian Lovers*
ca. 1960

## Plates

Plate 1:
*La Solana*
ca. 1928

Plate 2:
*Sam Borchard as a Young Man*
n.d.

Plate 3:
*Stuart Borchard*
ca. 1928

Plate 4:
*Llao Llao Trees*
1949

Plate 5:
*Three Carriages, Belgrano R* 1949

Plate 6:
*San Isidro Church, Buenos Aires* 1950

Plate 7:
*Ham Metzger*
1946

Plate 8:
*Suzie in Profile*
1946

Plate 9:
*Suzie, Mirror and Lamp*
1947

Plate 10:
*Model Knitting with Artist at Work*
n.d.

Plate 11:
*Model in Yellow Turban*
n.d.

Plate 12:
*Portrait of Jefferson Caffery*
1940

Plate 13:
*Bolivian Mother Nursing*
n.d.

Plate 14:
*Windsor Chair, Lytton Lodge on Lake Mooselookmeguntic*
ca. 1949

Plate 15:
*Jean Boley*
1946

Plate 16:
*Lucy Burrows*
1947

Plate 17:
*Ed and Jim*
ca. 1946

Plate 18:
*Jim Reading*
ca. 1947

Plate 19:
*Eva Borchard*
ca. 1943

Plate 20:
*Eruption*
1959

Plate 21:
*Molten Tide*
1958

Plate 22:
*Black, Yellow and Grey*
1959

Plate 23:
*Madonna and Child
Embracing*
ca. 1958

Plate 24:
*Madonna and Child
Enthroned*
n.d.

Plate 25:
*Cannes*
ca. 1964

Plate 26:
*Strasbourg*
ca. 1964

Plate 27:
*Anemones I*
1965

Plate 28:
*A Difficult Choice*
1966

Plate 29:
*Head of African Woman*
1969

Plate 30:
*Ethiopian Man*
1969

Plate 31:
*Washerwomen*
1966

Plate 32:
*Yellow and Blue Bouquet*
1968

Plate 33:
*Green Apples and Pears I*
1969

Plate 34:
*Indian Girl*
1965

Plate 35:
*Monumental Angkor
Figures*
ca. 1960

Plate 36:
*Angkor Roots (Temple)*
ca. 1960

Plate 37:
*Tribesman with
Umbrella*
1993

Plate 38:
*Spirit of the Crocodile
(New Guinea)*
1993

Plate 39:
*Status Symbol
(New Guinea)*
1994

Plate 40:
*Highlands New Guinea
Woman with Drum*
1993

Plate 41:
*Back View of Three
New Guinea Women*
1994

Plate 42:
*Storm Brewing on
Black Lakes, New Guinea*
1994

Plate 43:
*Black and White
Highlanders in Procession*
1993

Plate 44:
*Mourning (New Guinea)*
1993

Plate 45:
*Sunset on the Sepik,
New Guinea*
1994

Plate 46:
*Mud Man,
Burnt Orange Ground*
1993

Plate 47:
*Huli in Headdress,
New Guinea*
1994

Plate 48:
*Tribal Leader,
New Guinea*
1993

Plate 49:
*Hulis in Transit*
1993

Plate 50:
*Antarctica (Ten Penguins)*
1994

Plate 51:
*Ed in Antarctica*
1993

Plate 52:
*Distance (Antarctica)*
1993

Plate 53:
*Spirit of the Moai over
Easter Island*
1994

Plate 54:
*Moai in Quarry
(Easter Island)*
1994

Plate 55:
*Evanier Window*
1991

Plate 56:
*Winter View of Evanier
from Birches*
1992

Plate 57:
*Chandra in the Snow*
1993

Plate 58:
*Eva in Evanier*
1994

Plate 59:
*Eva Looking over
Her Shoulder*
ca. 1962

Plate 60:
*Eva and Chandra
(Mirror)*
1983

Plate 61:
*Chandra en peignoir*
1990

Plate 62:
*Self-Portrait with Art*
1994

Plate 63:
*Elizabeth*
1990

Plate 64:
*Elizabeth Modeling
Yellow Hat*
1991

Plate 65:
*Jimmy at Quogue
Pool Fence*
1985

Plate 66:
*Jimmy at Three with
Teddy Bear*
1985

Plate 67:
*At Water's Edge,
Quogue Beach*
1986

Plate 68:
*Boy Architects on
Quogue Beach*
1986

Plate 69:
*Tasseled Grass,
Quogue Refuge*
1987

Plate 70:
*White Crescent,
Turquoise Water*
1980

Plate 71:
*Anemones in Greece*
1978

Plate 72:
*Mallika in Red*
1993

Plate 73:
*Mallika in Costa Rica*
1993

Plate 74:
*Park Bench with Shadows*
1982

Plate 75:
*San Remo Reflections,
Central Park*
1994

Plate 76:
*Tulip Magnolia at the Frick*
1986

Plate 77:
*Park Snowscape,
Twisted Branches*
1992

Plate 78:
*Maple Tree in Fall,
Central Park*
1991

Plate 79:
*Park Snowscape
with Benches*
1992

Plate 80:
*Park Snowscape with Fir*
1992

Plate 81:
*Central Park Path
in Fall*
1991

Plate 82:
*Dog-Walker*
ca. 1979

Plate 83:
*Chatting at
59th Street Plaza*
1991

Plate 84:
*Baby Watching Musicians,
Central Park*
1993

Plate 85:
*School Class, Central Park
(Glancing Back)*
1992

Plate 86:
*Family and Tricycle
at the Pond, Central Park*
1993

Plate 87:
*Camels in Central Park*
1991

Plate 88:
Detail, *Bethesda Fountain
Panorama (Blue Banner)*
1991

Plate 89:
*Solitary Reader,
Boat House Café*
1991

Plate 90:
*Boat House Café,
Striped Umbrellas*
1993

Plate 91:
*Cyclist at New York
Marathon (Run Roger)*
1994

Plate 92:
Detail, *Tavern Terrace
Panorama, Central Park*
1993

Plate 93:
*Sunbathers on Grass,
Central Park*
1989

Plate 94:
*Willows, Central Park*
1991

Plate 95:
*Sunny Day in Park
with Fence*
1993

Plate 96:
*Chess in the Park*
ca. 1982

Plate 97:
*Falling Snow,
Central Park*
1985

Plate 98:
*Blue Carriage with
Brown Horse, Central Park*
1992

Plate 99:
*Red-Lined Carriage
with White Horse,
Central Park*
1992

Plate 100:
*White Carriage
Pulling Away*
ca. 1992

Plate 101:
*Winter Bridge with
Lamppost, Central Park*
n.d.

Plate 102:
*Tavern Terrace Panorama,
Central Park*
1993

Plate 103:
*Park Bench with
Young Child*
1982

Plate 104:
*Homeless Man
with Shopping Bag,
Central Park*
ca. 1991

Plate 105:
*Melancholy (Central Park)*
ca. 1990

Plate 106:
*Schemmy's Ice Cream Parlor,
Rhinebeck*
1994

Plate 107:
*House in Distance, Rhinebeck*
1994

Plate 108:
*Ghost Ballet*
1994

Plate 109:
*Pas de Deux
(Black Costumes)*
1994

Plate 110:
*Ruisdael Redux*
1994

Plate 111:
*Brueghel Redux*
1994